Acta Biotechnologica

0 · 1980 · Volume 0

Journal of microbial, biochemical and bioanalogous technology

Akademie-Verlag
Berlin

THEORIA CUM PRAXI

Instructions to Authors

1. Only original papers that have not previously been published will be accepted.
   Manuscripts may be submitted in English, German or Russian but always in Duplicate. The name of the institute (inclusive full address) from which the manuscript originates should be stated below the author's name(s). The authors are responsible for the contents of their contributions.

2. Original papers should not consist of more than 20 typewritten pages (2 lined-spacing), including references, tables and figures; short original communications may have a maximum of 6 typewritten pages.

3. Each paper should be preceded by a summary in a language different from the used for the paper.

4. Latin names of species as well as passages to be printed in italics for greater emphasis should be signed with waved line. Please use only units and symbols of the SI-system.

5. Tables may be used to effect a reduction of the text or to make assimilation easier. They should be numbered consecutively throughout and supplied with a brief heading. They shouldn't be written among the text but on seperate sheets.

6. Number and size of illustrations should be limited to the essential minimum, they should be numbered consecutively and brought on seperate sheets.
   Line drawings including graphs and diagrams should be drawn in black ink. Half-tone illustrations should be supplied as white glossy prints. Figure legends are to be typed in sequence on a seperate sheet.
   Authors name is to be written on to the backside of each sheet.

7. References, listed at the end of the contribution, should contain only works cited in the text. They should be numbered in the order in which they are at first mentioned in the text. Please give surnames and initials of all authors, name of journal abbreviated according to "Chemical Abstracts — List of Periodicals", volume number, year of publication, issue number or month, first page number; books are to be cited with full title, edition, volume number, page number, place of publication, publisher and year of publication.

8. Notes on the text may be given as footnotes on the relevant page.

9. Publication and 50 reprints are free of charge.
   Additional reprints may be ordered.

10. The author recieves 2 galley proofs for correction.
    They are send back to the chief editor
    (Dr. Dimter, DDR-7050 Leipzig, Permoserstr. 15)
    as soon as possible.

Register of volume 0 (1980) and 1 (1981)

Contents

Number 0. 1980

Number 1. 1981

Number 2. 1981

Number 3. 1981

Number 4. 1981

Register of authors

Acta Biotechnologica

Journal of microbial, biochemical and bioanalogous technology

Published at
Institute of Technical Chemistry of the Academy of Sciences of the G.D.R.; Leipzig
and
Institute of Enzymology and Technical Microbiology of the Ministry of Chemical Industry of the G.D.R.; Berlin
by
M. Ringpfeil, Leipzig and D. Meyer, Berlin

Number 0

1980

Volume 0

AKADEMIE-VERLAG · BERLIN

"Acta Biotechnologica" publishes reviews, original papers, short communications and reports out of the whole area of biotechnology.
The journal shall promote the foundation of biotechnology as a new, homogeneous scientific field. According to biotechnology are microbial technology, biochemical technology and technology of synthesyzing and applying of bioanalogous reaction systems.
The technological character of the journal is guarenteed thereby that microbial, biochemical, chemical and physical contributions must show definitely the technological relation.

Terms of subscription for the journal "Acta Biotechnologica"
Orders can be sent

- **in the DDR:** to the Postzeitungsvertrieb, to a book-shop, or to the Akademie-Verlag, DDR - 1080 Berlin, Leipziger Str. 3—4;
- **in the other socialist countries:** to a book-shop for foreign languages literature or to the competent news-distributing agency;
- **in the BRD and Berlin (West):** to a book-shop or to the wholesale distributing agency Kunst und Wissen, Erich Bieber, D - 7000 Stuttgart 1, Wilhelmstr. 4—6;
- **in Österreich:** to the Globus-Buchvertrieb, A - 1201 Wien, Höchstädtplatz 3;
- **in the other Western European countries:** to Kunst und Wissen, Erich Bieber GmbH, CH - 8008 Zürich, Dufourstraße 51;
- **in other countries:** to the international book- and journal-selling trade, to Buchexport, Volkseigener Außenhandelsbetrieb der DDR, DDR - 7010 Leipzig, Postfach 160, or to the Akademie-Verlag, DDR - 1080 Berlin, Leipziger Straße 3—4.

Acta Biotechnologica

Herausgeber: Institut für technische Chemie der AdW,
DDR - 7050 Leipzig, Permoserstr. 15
(Direktor: Prof. Dr. Manfred Ringpfeil) und
Institut für Enzymologie und Technische Mikrobiologie
DDR - 1017 Berlin; Alt-Stralau 62
(Direktor: Dr. Dietrich Meyer).

Verlag: Akademie-Verlag, DDR - 1080 Berlin, Leipziger Straße 3—4;
Fernruf: 2236221 und 2236229; Telex-Nr.: 114420; Bank: Staatsbank der DDR, Berlin, Konto-Nr.: 6836-26-20712.

Redaktion: Dr. Lothar Dimter (Chefredakteur),
Martina Bechstedt (Redakteur), DDR - 7050 Leipzig,
Permoserstr. 15; Tel.: 6861255.

Gesamtherstellung: VEB Druckhaus „Maxim Gorki", DDR - 7400 Altenburg.

Erscheinungsweise: Die Zeitschrift „Acta Biotechnologica" erscheint jährlich in einem Band mit 4 Heften. Bezugspreis eines Bandes 100,— M zuzüglich Versandspesen; Preis je Heft 25,— M.

Bestellnummer dieses Heftes: 1094/0/1.

Ag 451/84/80

Acta Biotechnologica 0 (1980) 0, 3

Foreword

W. Scheler

President of the Academy of Sciences of the G.D.R.

Science has always to be considered as both a product and an element of the development of productive forces. While in this process science itself undergoes characteristic changes, increasing differentiation as well as the formation of interdisciplinary fields are especially obvious. Technical implementation of new scientific knowledge yields immediate practical consequences. In a historical setting, we observe an empirico-pragmatic approach in the beginning, which, however, in course of time is more and more substituted by systematic and scientific principles. Thus, various technico-scientific disciplines have emerged out of physics and chemistry. They are of crucial importance both for the nature and efficiency of material production.

In agriculture and forestry, technical aspects and technology of biological production exercise an ever growing influence in terms of the latter's output. Industrialization of agricultural production has reached a high level. While further developing, it approaches to an increasing extent the borderlines of the genetic potentials of plants and animals. During the past few decades, microbiological production has taken an especially dynamic development. Its technology has scientifically been opened up. Thanks to the fundamental progress made in the fields of biochemistry, molecular and cell biology, genetics, immunology and others, new starting-points in technical microbiology as well as new possibilities of controlling and of technically mastering reproduction, growth and specific capacities in plant and animal production are offering themselves.

We are able to perform in-vitro cultivation of plant and animal cells. The techniques of genetic engineering render possible modification of the genetic material of procaryontic and eucaryontic cells. We continue to improve our understanding of mathematic modelling, automation and process control as applied to the development and the use of technologies in fermentation processes. This development has led and is still leading to the formation of biotechnology as a new scientific discipline. Both the contours and the inner structure of this field are outlining themselves with increasing clearness. It represents an essential link between biology, technology and national economy. Therefore, the edition of "Acta Biotechnologica" as a scientific journal dedicated to promote and support further development of biotechnology has to be welcomed very much on behalf of both the development of science and production.

Acta Biotechnologica 0 (1980) 0, 5—14

Рост дрожжей на поверхности плотной среды и некоторые проблемы ферментации

Я. Э. Блумберг, М. Е. Бекер, И.К. Бривкалне

Институт микробиологии им. Августа Кирхенштейна
АН Латвийской ССР
г. Рига, Клейсти

Summary

A mathematic model of microorganism colony growth on solid media was worked out, the substrate diffusion taken into account.
The sample being Sacch. cerevisiae yeast correspondence of experimental and calculated data was demonstrated. Sacch. cerevisiae yeast specific growth rates on various substrates were determined. Some parameter effect on colony growth kinetics on a solid medium surface was shown. Two phases of limitation were detected in microorganism colony development — kinetical and diffusional ones. On the basis of calculations, theoretical suppositions and experimental data a conclusion was made that solid fermentation required the finest grinding possible, taking into account growth limitation by diffusion. Besides, dense seeding of all the medium surface with cells was found important.

В последнее десятилетие в микробиологической промышленности укоренилось представление, что более прогрессивным и перспективным методом культивирования микроорганизмов по сравнению с поверхностным является глубинная ферментация. Действительно, современное производство антибиотиков, аминокислот, витаминов, ферментов, спиртов, растворителей и других продуктов, а также крупнотоннажное производство дрожжей, основано на глубинном процессе.
Основным преимуществом глубинной ферментации является то, что целевой продукт — внеклеточное вещество или биомасса — легче отделимо от субстрата, гомогенно по всему объему ферментируемой массы, процесс протекает равномернее, его легко контролировать и управлять им, все технологические операции хорошо поддаются механизации и автоматизации [1, 2]. В то же время необходимо отметить, что при производстве, например, лимонной кислоты, с глубинным методом успешно конкурирует поверхностный, главным образом, по технологическим соображениями [3]. По причине необходимости применения сжатого воздуха для аэрации среды и механических мешалок для перемешивания культуральной жидкости при глубинном процессе требуются большие затраты энергии — 3 и более кВт.ч на 1 кг кислоты. Долголетний опыт использования поверхностной ферментации при производстве лимонной кислоты на Экспериментальном заводе биохимических препаратов Института микробиологии им. Августа Кирхенштейна АН Латв. ССР показывает возможность ее получения при удельных энергозатратах на стадии ферментации 0,3—0,4 кВт.ч на 1 кг кислоты. Таким образом, на единицу продукции при глубинной

ферментации лимонной кислоты расходуется на порядок больше энергии по сравнению с поверхностной.

Лимонную кислоту при поверхностном методе культивирования *Asp. niger* получают в жидкой среде в кюветах под пленкой грибного мицелия. В этом случае физиология развития продуцента незначительно отличается от развития при глубинном росте, если не считать того, что кислород воздуха поступает в клетки непосредственно из газовой фазы через тонкий пограничный слой жидкой среды. Более отличительные условия существуют для микроорганизмов, если они развиваются на плотной питательной среде, что очень широко распространено в природе. В этом случае питательные вещества поступают в клетки путем диффузии, и фактором, лимитирующим рост, может стать скорость диффузии какого-либо субстрата.

Культивирование микроорганизмов на плотных средах привлекает внимание исследователей в связи с получением гидролаз углеводов и богатой белком микробной биомассы из природного растительного сырья. В промышленных условиях широко известно производство ,,грибного солода" при помощи микроскопических грибов из рода *Aspergillus* на средах из пшеничных отрубей и других растительных крахмалосодержащих материалов [2, 4, 5].

Проведена ферментация рисовых отрубей при помощи *Asp. oryzae* и получен кормовой продукт, замена которым неферментированных отрубей в рационе цыплят, дало увеличение живой массы на 16% [6].

В последнее время появились работы по вопросу увеличения слоя среды до 15 до 20 см (по сравнению с 2—3 см при производстве ,,грибного солода"). Необходимым условием являются приготовление гранул твердой среды одинакового размера и принудительная циркуляция воздуха через уплотненный слой [7]. Например, после 30 часов ферментации по данному методу из 100 кг муки маниока получают 76 кг сухого продукта, содержащего 13,5 кг белка (по Лоури) и 19,8 кг углеводов. Коэффициент конверсии углеводов в белок составляет 17%, что соответствует результатам, полученным с данным штаммом в условиях глубинной ферментации. Авторы утверждают, что такой способ культивирования спорообразующих организмов на плотном слое гранул с мелкими порами исключает спорообразование, что имеет существенное значение для получения кормовых продуктов.

Ферментацию на плотных средах начинают использовать также при ферментации целлюлозосодержащих субстатов [8].

Эти примеры свидетельствуют о возрастающем интересе к поверхностной ферментации как к методу менее энергоемкому, чем глубинная ферментация. Необходимо также отметить, что в случае использования плотных сред для получения технических ферментационных препаратов типа концентратов, а также белковоуглеводных концентратов кормового назначения, этот метод дает большую экономию энергии на заключительной стадии, т. е. при обезвоживании продуктов ферментации. По ориентировочным расчетам тепловые затраты при обезвоживании продуктов ферментации на плотных средах в 3 раза ниже, чем при полном обезвоживании продуктов глубинной ферментации. К преимуществам поверхностной фермертации следует также отнести упрощенное аппаратурное оформление процесса.

Несмотря на то, что поверхностная ферментация в микробиологическом производстве известна давно и широко применяется также в лабораторной практике, теория развития микробных популяций в условиях роста на плотных средах разработана значительно слабее, чем соответствующая теория для глубинного роста. Это касается как кинетики роста популяции, так и физиолого-биохимических аспектов продуцентов, а также вопросов массопередачи основных компонентов питания из плотной среды в клетки микроорганизмов.

Некоторые обобщения вопросов роста колоний бактерий и грибов на поверхности плотных (агаровых) сред, сделано Пертом (1978). Автор предполагает, что в начале, когда внесено небольшое количество клеток микроорганизмов, размножение протекает так, что все клетки в одинаковой степени участвуют в увеличении популяции. Рост всей популяции идет с максимальной экспоненциальной скоростью до тех пор, пока концентрация источников питания остается значительно выше константы насыщения. Согласно Перту рост колонии в основном зависит от экспоненциального роста, ограниченного периферической зоной с постоянной шириной (w). Следует вывод, что после экспоненциальной фазы рост идет по линейному закону:

$$r = \mu wt + r_0 \tag{1}$$

где r — радиус колонии

t — время

μ — удельная скорость роста культуры

У бактерий и нитчатых грибов радиальные скорости роста различные. В обоих случаях они зависят от концентрации питательных субстратов. При этом экспериментально обнаружено,что существует пороговая концентрация, выше которой начинается рост колонии.

По мнению других авторов [9], рост колонии бактерий описывается регрессионным уравнением:

$$r = -A + Bt - Ct^2 \tag{2}$$

где A, B, C;— эмпирические коэффициенты.

При исследовании роста колоний, по мнению Перта, основное внимание должно быть уделено изучению зависимости радиальной скорости роста клеток от удельной. Хотя дрожжи имеют широкое промышленное применение, в литературе очень мало данных о росте дрожжей на поверхности плотных сред.

Методика

Культура дрожжей — Для исследования роста колоний на плотных средах использовали дрожжи *Sacch. cerevisiae* **14**, полученные из коллекции Института микробиологии им. Августа Кирхенштейна АН Латв. ССР. Посевным материалом служила 48-часовая культура, выращенная на сусло-агаре.

Питательные среды — Источники углерода — меласса, глюкоза, этанол. Остальные компоненты среды в г/л: $MgSO_4$ — 0,7, NaCl — 0,5, $(NH_4)_2SO_4$ — 7,54, KH_2PO_4 — 1,0, K_2HPO_4 — 0,13, Бакто-агар — 20.

Перед стерилизацией дважды фильтровали.

Выращивание колоний дрожжей осуществляли в чашках Петри с диаметром 50—90 мм. Перед высевом дрожжевой культуры чашки с агаровой средой подсушили при 25°С в течение 15 минут [10]. Высев дрожжей проводился нанесением на чашки по 0,1 мл суспензии дрожжей или внесением в центр чашки одной клетки дрожжей (контролировалось микроскопически). Выращивание проводили в термостате при 30°С.

Измерение параметров колоний — Диаметр колоний измерялся с помощью проекционного микроскопа *Pictoval* [11]. Для измерения формы колонии их заливали 2%-ным раствором Бакто-агара и делались поперечные разрезы колоний, которые также измеряли с помощью проекционного микроскопа. Для ограничения радиального роста использовались пластмассовые цилиндрики.

Математическая модель

При построении математической модели роста колонии микроорганизмов на поверхности плотной среды (рис. 1) сделаны следующие допущения:

1) поток субстрата в плотной среде происходит за счет диффузии;
2) коэффициент диффузии (Д) не зависит от концентрации субстрата;
3) концентрация субстрата в какой-либо точке колонии зависит только от координат x и z и равна концентрации субстрата (S_A) на поверхности плотной среды в данной точке (см. рис. 1);
4) диффузия субстрата внутри колонии не учтывается;

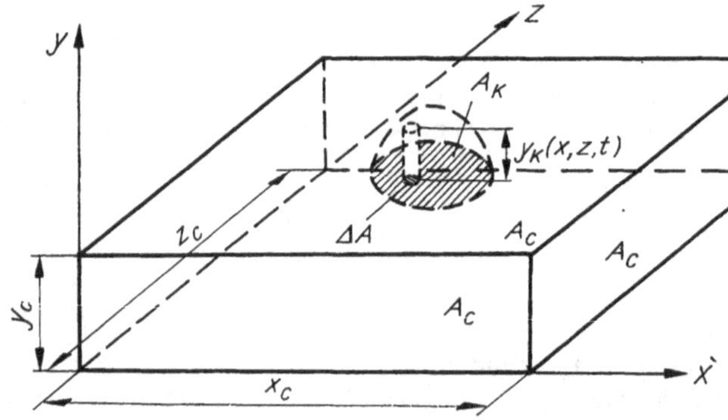

Рис. 1. Схематическая модель колонии на поверхности плотной среды

x_c; y_c; z_c — размеры плотной среды
y_K — высота колонии в точке x
A_c — свободная поверхность среды
A_K — основание колонии

5) удельная скорость роста какой-либо области (ΔV) колонии зависит только от максимальной удельной скорости роста ($\mu_{\max}$) и концентрации субстрата в этой области колонии по закономерности

$$\frac{d\Delta V}{dt} = \Delta V \cdot \mu_{\max} \cdot \frac{S_A(x;\, z;\, t;)}{K_S + S_A(x;\, z;\, t)} \tag{3}$$

6) поток субстрата через какой-либо участок колонии с основанием (ΔA) и высотой y_K, который используется для роста колонии, пропорционален количеству биомассы на единицу объема (ϱ), скорости роста данного участка колонии и обратно пропорционален экономическому коэффициенту (Y)

$$\Delta j_- = \frac{\varrho}{Y} \cdot \frac{d\Delta V}{dt} \tag{4}$$

подставляя уравнение 3, получаем

$$\Delta j_- = \frac{\varrho \mu_{\max} \Delta A S_A \cdot y_K}{Y(K_S + S_A)} \tag{5}$$

7) диффузионный приток субстрата к колонии определяется первым законом Фика

$$\Delta j_{+} = \Delta A \cdot D \left(\frac{\delta S}{\delta y}\right)_{A_K} \tag{6}$$

8) потери субстрата через свободную от колонии поверхность плотной среды равен нулю, т. е.

$$D \left(\frac{\partial S}{\partial y}\right)_{A_c} = 0 \tag{7}$$

9) баланс притока и оттока субстрата на поверхности среды

$$D \left(\frac{\partial S}{\partial y}\right)_{A_K} = \frac{\varrho \mu_{\max} \cdot y_K(x; z; t;)}{Y(K_S + S_A)} \tag{8}$$

10) концентрация субстрата на поверхности плотной среды определяется вторым законом Фика и начальным распределением субстрата в среде

$$\frac{dS}{dt} = D\nabla^2 S; \quad S_0 = S(x; y; z) \qquad \text{при } t = 0 \tag{9}$$

Таким образом, уравнения (3), (7), (8) и (9) составляют математическую модель роста колоний микроорганизмов на поверхности плотных сред.

Рассмотрим частный случай модели, при условии, что диффузия происходит только в плоскости x и y и предположим, что колония растет только по оси y, а концентрация субстрата зависит только от времени. При сделанных допущениях модель имеет следующий вид:

$$\frac{dy_K}{dt} = y_K(t)\, \mu_{\max} \frac{S_A(t)}{K_S + S_A(t)}$$

$$D \left(\frac{\partial S}{\partial y}\right)_{A_K} = \frac{\varrho y_K(t)\, \mu_{\max} \cdot S_A(t)}{Y[K_S + S_A(t)]}; \qquad D \left(\frac{\partial S}{\partial y}\right)_{A_c} = 0 \tag{10}$$

$$\frac{dS}{dt} = D \left(\frac{\partial^2 S}{\partial x^2} + \frac{\partial^2 S}{\partial y^2}\right)$$

Данная задача решалась на ЭВМ при следующих величинах параметров: $y_{KO} = 10$ мкм, $x_{KO} = 0{,}5$ мм, $Д = 3 \cdot 10^{-6}$ см³/сек, $S_0 = 2\%$, $K_S = 0{,}005$ г/см³, $\varrho = 0{,}1$ г/см³, $Y = 0{,}5$, $y_c = 5$ мм, $X_3 = 5$ мм, $\mu_{\max} = 0{,}36$ час$^{-1}$. Рассчитали (рис. 2) кинетику роста колонии, концентрацию субстрата на поверхности среды, логарифм скорости роста. На рис. 3 показана расчетная кинетика роста колонии при условии, что $x_{KO} = x_c = 0{,}5$ см. Из данных результатов видно, что в развитии колонии можно выявить по крайней мере две фазы роста: 1-я фаза — логарифм скорости роста пропорционален времени, 2-я фаза — логарифм скорости роста обратно пропорционален времени. Первая фаза роста характеризуется экспоненциальным ростом колонии. Скорость роста во время этой фазы лимитируется величиной удельной скорости роста клеток. Таким образом, эта фаза соответствует кинетическому лимитированию роста. Во второй фазе развития, концентрация субстрата в колонии близка к нулю, и скорость потока субстрата определяется скоростью диффузии, т. е., эта фаза соответствует лимитированию роста массопередачей.

Если концентрация диффундирующего вещества на поверхности плоской пластинки равна нулю, то зависимость концентрации от времени можно выразить в явном виде (12):

$$S = \frac{4S_0}{\pi} \sum_{n=1}^{\infty} \frac{(-1)^{n-1}}{2n-1} e^{-\left(\frac{2n-1}{2y_c}\pi\right) Dt} \cos \frac{2n-1}{2y_c} \pi y \qquad (11)$$

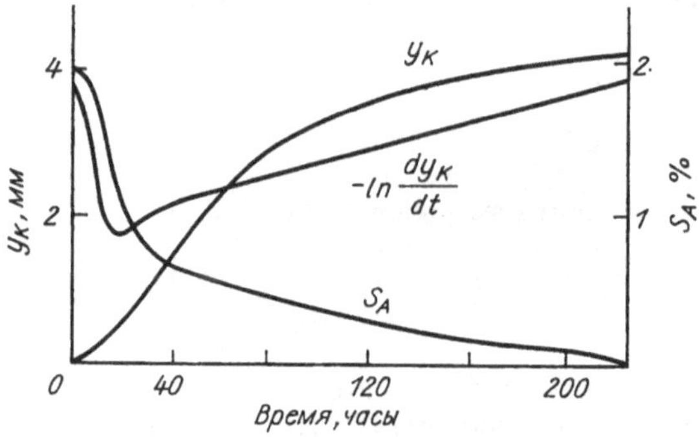

Рис. 2. Расчетная кинетика вертикального роста колонии микроорганизмов
y_K — высота колонии
S_A — концентрация субстрата на поверхности среды под колонией,
$x_K = 0{,}1\ x_c$

Учитывая первый закон Фика и уравнение (11), и взяв первый член разложения в первом приближении при больших t и $x_K = x_c$, получаем:

$$\ln \frac{dy_K}{dt} = \frac{\pi^2 D}{4{y_c}^2} \cdot t \qquad (12)$$

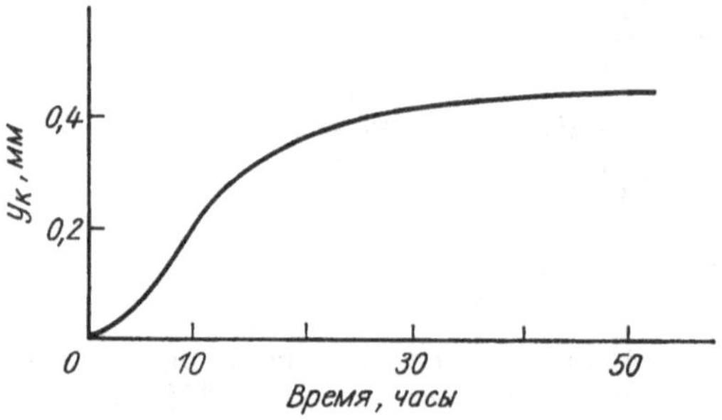

Рис. 3. Расчетная кинетика вертикального роста колонии микроорганизмов
y_K — высота колонии, $x_K = x_c$

Если колония растет в центре на поверхности цилиндрической плотной среды с радиусом R и при $R \gg y_c$ и $x_K \sim y_c$, получаем

$$\ln \frac{dx_K}{dt} \approx \frac{\pi^2 D}{4R^2} t \qquad (13)$$

Из уравнения (11) следует, что скорость диффузии субстрата с поверхности плотной среды прямо пропорциональна начальной концентрации субстрата и экспоненциально падает. Учитывая, что клетки потребляют часть субстрата в процессах, не связанных прямо с их ростом, можно определить в первом приближении начальную концентрацию субстрата, ниже которой рост клеток на поверхности плотной среды не происходит, так как весь поток субстрата потребляется в процессах, не связанных с ростом. При $x_{KO} = x_c$, получаем:

$$S_0 > \frac{y_{KO} \cdot y_c \cdot \varrho}{2DY} \cdot a \tag{14}$$

где a — удельная скорость поддержания

Это совпадает с данными других авторов [13, 10], показавших, что пороговая концентрация увеличивается с увеличением парциального давления кислорода, причем быстро повышается потребление субстрата в процессах, не связанных с ростом клеток.

Продуктивность (Q) ферментации на поверхности плотной среды можно представить следующим уравнением:

$$Q \approx \frac{V_K - V_{KO}}{t \cdot V_C} \tag{15}$$

где V_K — объем колонии
V_C — объем среды

Максимальная продуктивность ферментации достигается, если клетки растут с постоянной удельной скоростью роста (μ), т. е. во время первой фазы (t_1) и, если $y_K = \mathrm{f}(t)$, то

$$Q = \frac{y_{KO}}{y_c} \cdot \frac{(\mathrm{e}^{\mu t_1} - 1)}{t_1} \tag{16}$$

Для оценки длительности первой фазы роста определяем момент, когда средний поток диффузии с поверхности, на которой $S_A = 0$ и средний поток субстрата, потребляемого колонией при экспоненциальном росте, равны:

$$\frac{\varrho y_{KO} \mu}{Yt} \int_0^{t_1} \mathrm{e}^{\mu\tau}\, d\tau = \frac{2S_0 D}{y_c t} \int_0^{t_1} \mathrm{e}^{-\frac{\pi^2 D\tau}{4y_c^2}}\, d\tau \tag{17}$$

Из уравнения (14) при $\mathrm{e}^{-\frac{\pi^2 D\tau}{4y^2c}} \ll 1$ получаем

$$t_1 \doteq \frac{1}{\mu} \ln\left(1 + \frac{8S_0 y_c Y}{\pi^2 \varrho y_{KO}}\right) \tag{18}$$

подставив уравнения (16) и (18), получаем

$$Q = \frac{S_0 \mu Y 8}{\pi^2 \varrho} \left[\ln\left(1 + \frac{8S_0 y_c Y}{\pi^2 \varrho y_{KO}}\right)\right] \tag{19}$$

Следовательно, продуктивность ферментации на плотных средах попорцопнальна удельной скорости роста микроорганизмов, начальной концентрации субстрата и экономическому коэффициенту.

Результаты

На рис. 4 показана кинетика радиального и вертикального роста колоний дрожжей при диаметре среды 90 мм и толщине 5 мм. Из полученных результатов видно, что в развитии колонии дрожжей на поверхности агаровой среды наблюдаются две отличительные по характеру роста фазы.

На рис. 5 показаны поперечные разрезы колоний дрожжей в разные моменты их развития. Видно, что в течение первых 50 часов форма колонии существенно не меняется. Это значит, что удельные скорости радиального и вертикального роста в

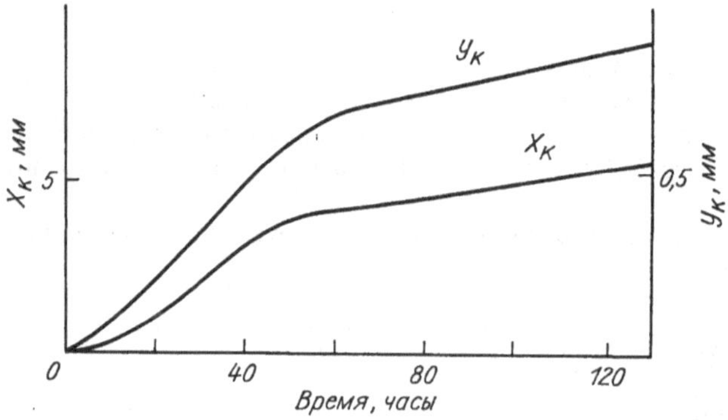

Рис. 4. Кинетика диаметра колонии (x_K) и высоты колонии (y_K) при росте дрожжей на поверхности 2% мелассной среды

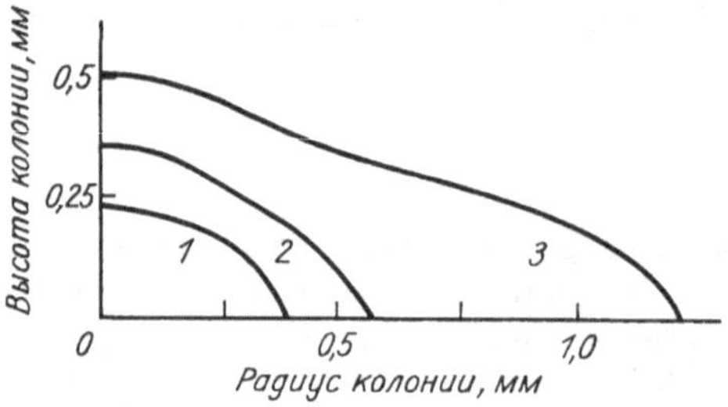

Рис. 5. Половина поперечного разреза колонии дрожжей *Sacch. cerevisiae* 14 на разных стадиях роста

1 — 27 час
2 — 47 час
3 — 119 час (масштаб 1:2)

этот период времени равны. По логарифму от радиальной скорости роста можно определить его удельную скорость. Если она равна удельной скорости вертикального роста, можно определить также удельную скорость роста дрожжей на поверхности плотной среды.

Табл. 1. Удельная скорость роста дрожжей *Sacch. cerevisiae* 14, выращенных на плотных средах с различными субстратами

Меласса %	μ	Глюкоза %	μ	Этанол %	μ
0,05	0,25	0,05	0,33	1,5	0,23
0,5	0,36	0,5	0,28		
2	0,41	2	0,43		
		10	0,07		
		15	0,09		
		20	0,06		

В таблице 1 приведены удельные скорости ($\pm 20\%$) роста дрожжей на поверхности плотных сред, содержащих некоторые субстраты в разных концентрациях. Сравнение полученных данных по росту дрожжей на этих же субстратах в глубинных условиях с литературными данными показывает, что удельные скорости роста дрожжей в обоих случаях существенно не отличаются [14, 15, 16].

Производная по времени от логарифма радиальной скорости роста колоний во второй фазе развития на плотной среде диаметром 50 мм и толщиной 1 мм, находится в пределах от 0,014 до 0,027 час$^{-1}$, а при выращивании дрожжей на среде диаметром 90 мм — 0,006—0,008 час$^{-1}$.

Если принять коэффициент диффузии глюкозы в агаровой среде при $T = 30°$ за $5{,}5 \cdot 10^{-6}$ см^2/сек [9], то по формуле (13) получаем 0,01 час$^{-1}$ в первом случае и 0,003 час$^{-1}$ — во втором. Отсюда видно, что расчетные и экспериментальные результаты совпадают.

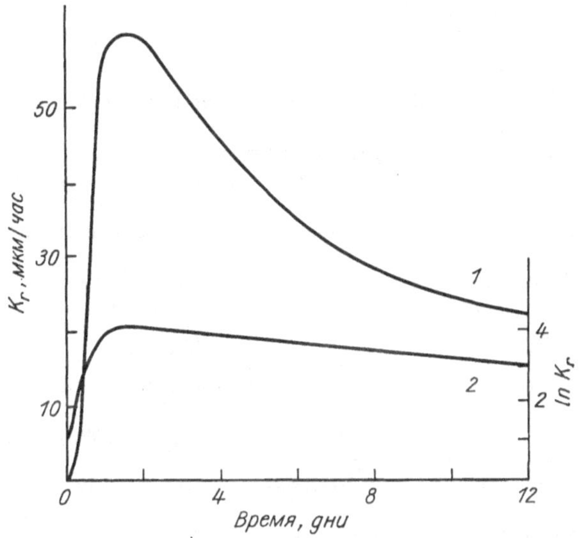

Рис. 6. Кинетика роста дрожжей *Sacch. cerevisiae* 14 на чашках Петри с диаметром 9 см

1 — радиальная скорость роста
2 — логарифм радиальной скорости роста
2%-ная мелассная среда; толщина питательной среды — 5 мм

На рис. 6 показаны кинетика радиальной скорости и логарифм радиальной скорости роста колонии дрожжей при росте на 2%-ной мелассной среде диаметром 90 мм и толщиной 5 мм. Из рис. 6 более отчетливо, чем из рис. 4 видны две характерные фазы роста. Максимальная радиальная скорость роста достигается к концу 48-ого часа развития и составляет 60 мкм/час. По уравнению (18), при $\mu = 0{,}3$ час$^{-1}$, $Y = 0{,}5$, $\varrho = 0{,}1$ г/см^3, получаем $t_1 = 20$ часов. Для получения более точной величины длительности первой фазы роста необходимо решить уравнения (3), (7), (8) и (9). Изучение влияния плотности засева свободной поверхности питательной среды показывает, что при засеве всей поверхности 2%-ной мелассной среды диаметром 90 мм и толщиной 5 мм прирост биомассы в течение 24 часов составляет 24 мг (85% от теоретически возможного при экономическом коэффициенте 0,5). При засеве одной сотой части поверхности среды прирост биомассы за этот же период времени составляет 3 мг или 7,5% от максимального.

Расчеты на ЭВМ показывают, что в первом случае прирост биомассы должен составить 72% от максимально возможного и во втором случае — 16%.
Резюмируя выше изложенное, можно констатировать, что при ферментации на плотных средах скорость роста во второй фазе лимитируется диффузией субстрата и в целях увеличения продуктивности системы для роста микроорганизмов целесообразно использовать частицы минимальных размеров с максимально допустимой (не снижающей μ) концентрацией субстрата, обеспечив при этом плотный засев поверхности биомассой.
Культивирование микроорганизмов на поверхности плотных сред, ввиду заметно сниженных энергозатрат, является перспективным для получения микробного белка и гидролаз полисахаридов из крахмало- и целлюлозосодержащих субстратов.

Eingegangen: 28. 12. 1979

Литература

[1] Айба Ш., Хемфри А., Миллис Н.: Биохимическая технология и аппаратура, ,,Пищевая промышленность'' М., 1975, 287 с.
[2] Бекер М.: Введение в биотехнологию, ,,Пищевая промышленность'', М., 1978, 231 с.
[3] Карклиньш Р. Я., Пробок А. К.: Биосинтез органических кислот. Рига, ,,Зинатне'', 1979, 198 с.
[4] Грачева И. М.: Технология ферментных препаратов. М. ,,Пищевая промышленность'', 1975, 392 с.
[5] Яровенко В. Л., Калунянц К. А., Ломер Т. И.: Производство ферментных препаратов из грибов и бактерий. М. ,,Пищевая промышленность'', 1970, 444 с.
[6] Miglietta H. F., Marcipar A. J.: IL RISO (Milan), 16 (1977) 4, 331.
[7] Raimbault M., Germon J. C.: France N 234429, Cl. A. 2371/18, A 23 L 1/218, 1977.
[8] Лобанок А. Г., Бабицкая В. Г.: Микробиологический синтез белка на целлюлозе. Минск, ,,Наука и техника'', 1976, 230 с.
[9] Rieck V. T., Palumbo S. A., Witter L. D.: J. Gen. Microbiol. 74 (1973) S. 1.
[10] Pirt S. I.: J. Gen. Microbiol. 47 (1967) S. 181.
[11] Перт С. Дж.: Основы культивирования микроорганизмов и клеток. Москва ,,Мир'' 1978, 332 с.
[12] Jacob M.: "J. Wiley and Sons", New York, "Chapman and Hall", London, 1950, 758.
[13] MacLeannan D. Y., Ousby I. C., Vasey R. B., Cooton N. T.: J. Gen. Microbiol. 69 (1971) S. 395.
[14] Beck C., von Meyenburg M. K.: J. Bacteriol. 96 (1968) S. 479.
[15] Fernandes M. J., Pou I., Gonzales P., Velac C., Gorrido J.: Symp. of Yeast, Vienna, 1974, part 1, 155.
[16] Nagai S., Nishizawa Y., Aiba S.: J. Gen. Appl. Microbiol. 19 (1973) S. 221.

Acta Biotechnologica 0 (1980) 0, 15—20

Aufnahme von Kohlenwasserstoffen durch Hefen (Teil I)

K. SATTLER, L. WÜNSCHE

Akademie der Wissenschaften der DDR
Institut für technische Chemie Leipzig
7050 Leipzig, Permoserstr. 15

Summary

The uptake of hydrocarbons by yeasts is measured by gravimetric determination of intracellular accumulated N-free compounds as well as by detection of tritium-labeled hydrocarbons. The aim of the investigations is to compare the uptake of n-alcanes and aromatics with respect to the conditions of yeast fermentation on gas oil. Aromatics in comparison to n-alcanes are taken up in very little amounts. The uptake of n-alcanes is dependent on the chain length. According to increasing chain length in the range of $C_{10}-C_{16}$ the velocity of uptake is increased whereas the optimal pH-value is decreased.
In the presence of dimethylsulfoxide the uptake of n-alcanes is markedly increased whereas the uptake of aromatics is decreased. A mechanism of hydrocarbon uptake is discussed characterized by obligate participation of metabolic active cells and hydrocarbon selectivity.

Einleitung

Als Maß für die Aufnahme von Kohlenwasserstoffen wurde sowohl die intrazelluläre Akkumulation wie der Nachweis markierter Verbindungen in der Zelle herangezogen. Auf der Grundlage einer einfachen Übersichtsmethodik wurde die Aufnahme in Abhängigkeit von der Struktur der Kohlenwasserstoffe untersucht unter besonderer Berücksichtigung der Unterschiede zwischen der n-Alkan- und Aromatenfraktion des Dieselkraftstoffs. N-Alkane werden mit zunehmender Kettenlänge im Bereich $C_{10}-C_{16}$ mit steigender Geschwindigkeit aufgenommen. Für die Aufnahme individueller n-Alkane wurden pH-Optima ermittelt, die sich mit zunehmender Kettenlänge nach dem sauren Bereich hin verschieben. So liegt das Optimum für C_{10} bei pH 8 und für C_{16} bei pH 5. Die Aromatenfraktion wird im Vergleich zu n-Alkanen nur in geringsten Mengen aufgenommen. Zur intrazellulären Akkumulation kommt es nicht. Dimethylsulfoxid wirkt auf die Aufnahme eines Gemisches von n-Alkanen und Aromaten unterschiedlich. Die Aufnahme der n-Alkane wird beschleunigt, die der Aromaten dagegen verlangsamt. Es werden Schlußfolgerungen auf einen Aufnahmemechanismus gezogen, der von aktiver Stoffwechseltätigkeit abhängig ist und selektiv wirkt.
Die Umsetzung von Kohlenwasserstoffen mit Hefen ist sowohl im Hinblick auf die Gewinnung von Futtereiweiß als auch durch die Erzeugung von Zitronensäure von praktischer Bedeutung. Das anhaltende Interesse an dieser Organismengruppe und an dieser Kohlenstoffquelle hat in den zurückliegenden zwei Jahrzehnten zu einer vielseitigen und intensiven biologischen und technologischen Bearbeitung geführt, die für

das Problem der Aufnahme von Kohlenwasserstoffen in die Zelle noch nicht abgeschlossen ist. Das Ziel der Untersuchungen besteht vor allem darin, die Substrataufnahme bei der Kohlenwasserstoffverhefung als limitierenden Faktor für Wachstum oder Produktakkumulation zu charakterisieren und nach Möglichkeit zu eliminieren und zwar unter den besonderen Bedingungen der gleichzeitigen Anwesenheit von assimilierbaren und nichtassimilierbaren Kohlenwasserstoffen. Eng damit verbunden ist die generelle Frage nach dem Mechanismus der Aufnahme solcher lipophiler Substrate durch Mikroorganismen.

Material und Methode

Untersucht wurden Hefen der Stammsammlung des Institutes für technische Chemie, Leipzig, bevorzugt die Stämme *Candida lipolytica* 43 und *Lodderomyces elongisporus* H. Als Maß für die Aufnahme der n-Alkane wurde die intrazelluläre Akkumulation N-freier Substanz (NFS) unter Stickstoffmangelbedingungen bei intensiver Belüftung innerhalb von 24 h herangezogen. Die Bestimmung erfolgte gravimetrisch.

Die Vorkultivierung wurde auf einem Mineralsalzmedium mit 2% Glucose durchgeführt. Das Versuchsmaterial wurde durch mehrfaches Waschen in Phosphatpuffer (nach *Sörensen*) gereinigt, auf eine Zellkonzentration von 3—4 g/l ($2 \cdot 10^8$ Zellen) eingestellt und Reste der C-Quelle durch zweistündige intensive Belüftung ausgezehrt.

Als C-Quelle wurde *Mepasin* verwendet, ein n-Alkangemisch im Kettenlängenbereich C_{12}—C_{18} (mit geringen Anteilen C_{10}, C_{11} u. C_{20}—C_{22}) oder individuelle n-Alkane, in der Regel in Konzentrationen von 10 g/l.

Gravimetrische Bestimmung: Zusatz von 1—2 ml 5%iger Dispersallösung und 50 ml Aceton, kurzes kräftiges Aufschütteln und Überführung in gewichtskonstante G4-Fritten.

Danach dreimaliges Absaugen mit Aceton bis zur Trockne und Nachwaschen mit 20 ml HCl-saurem Wasser. In den Fritten verbleibt die Hefe als weißes Pulver, frei von äußerlich anhaftenden Kohlenwasserstoffen. Trocknung bei 150 °C im Trockenschrank bis zur Gewichtskonstanz (18 h).

Drei Parallelbestimmungen, Abweichungen der Hefetrockensubstanz (HTS) maximal ±5%. pH-Verschiebung während der Versuchsdauer maximal 0,5 zum sauren Bereich hin.

Die Aufnahme von Kohlenwasserstoffen gilt als positiv, wenn eine Gewichtszunahme von mindestens 10% nachgewiesen wird. Die intrazellulär akkumulierte NFS besteht zu etwa 40% aus Lipiden und zu 60% aus Kohlenhydraten.

15—25% sind unveränderte Kohlenwasserstoffe. Das Verhältnis Lipide zu Kohlenhydraten verschiebt sich bei kürzeren Versuchszeiten zugunsten der Lipide.

Zur Untersuchung der Aufnahme und des Einflusses n-alkanfreier Erdöldestillate wurden herangezogen:

— Dieselkraftstoff, entparaffiniert
— Dieselkraftstoff, entparaffiniert und homogen tritiummarkiert
— Dieselkraftstoff, entparaffiniert und fraktioniert in:

- Fraktion Isoparaffine und Naphthene
- Aromaten
- Heterozyklen, besonders schwefelhaltige Verbindungen.

Dr. G. UHLENHUT (Zentralinstitut für Isotopen und Strahlentechnik Leipzig) danken wir für die Darstellung der tritiummarkierten Verbindungen und die Durchführung der Aktivitätsmessungen.

Ergebnisse

1. Einfluß der Kettenlänge von n-Alkanen auf die NFS-Akkumulation

Die Versuche wurden mit *Candida lipolytica* 43 im Kettenlängenbereich C_8—C_{24} bei dem in der Fermentation allgemein angewandten pH-Wert von 5 durchgeführt.
Octan wird nicht akkumuliert, aber zwischen Decan und Hexadecan wird mit steigendem Molekulargewicht eine zunehmende NFS-Akkumulation beobachtet.
Das Maximum liegt bei Hexadecan und Octadecan, während mit Eicosan und Tetracosan nur sehr niedrige Werte erreicht werden. Als Ursache kommt in Betracht, daß diese Substanzen bei der Versuchstemperatur von 30 °C bereits fest vorliegen (s. Abb. 1).

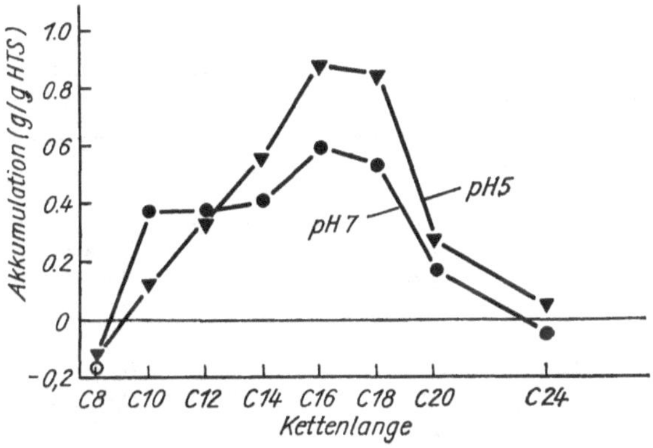

Abb. 1. NFS-Akkumulation von *Candida lipolytica* 43 in Abhängigkeit von der Kettenlänge des Substrates und dem pH des Mediums

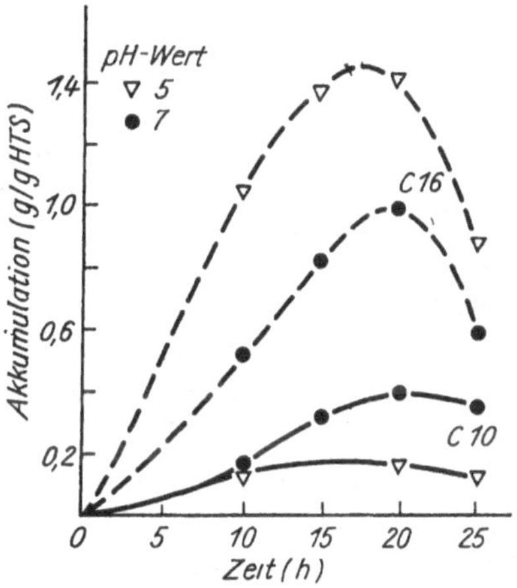

Abb. 2. NFS-Akkumulation bei *Lodderomyces elongisporus* H in Abhängigkeit von der Kettenlänge des Substrates und dem pH

Untersucht wurden auch die Fettsäuren der Kettenlänge C_{10}—C_{18}. Für Decansäure konnte keine Akkumulation nachgewiesen werden. Von der Dodecansäure an steigt die Akkumulation mit der Kettenlänge. Das Maximum wird bei Tetradecansäure erreicht. Hexadecansäure und Octadecansäure sind bei der Versuchstemperatur bereits fest.

2. Einfluß des pH-Wertes auf die NFS-Akkumulation in Abhängigkeit von der Kettenlänge der n-Alkane

Die auf n-Alkanen verschiedener Kettenlänge akkumulierte NFS-Menge ist deutlich vom pH-Wert abhängig. Für individuelle n-Alkane ergibt sich ein pH-Optimum der Akkumulation, das sich mit zunehmender Kettenlänge immer mehr in den sauren Bereich verschiebt (s. Abb. 1 u. 2).

3. Aufnahme nichtassimilierbarer Kohlenwasserstoffe

Bei der Verhefung von Erdöldestillaten bestehen die an die Hefezelle gelangenden Kohlenwasserstoffe zu 80—90% aus unverwertbaren Verbindungen und nur zu 10—20% aus den assimilierbaren n-Alkanen. Unsere Untersuchung war auf die Frage gerichtet, ob dieses Gemisch gleichmäßig in die Zelle gelangt oder selektiv aufgenommen wird und wie sich die Komponenten dabei gegenseitig beeinflussen.
Die Versuche wurden mit *Lodderomyces elongisporus* H durchgeführt.

Wird entparaffinierter tritiummarkierter Dieselkraftstoff als C-Quelle vorgegeben, so beträgt die intrazelluläre Akkumulation nur 0,033% des angebotenen Kohlenwasserstoffes.
Wird als Substrat ein dem Original-Dieselkraftstoff entsprechendes Modellgemisch aus 4 Teilen markiertem entparaffinierten Dieselkraftstoff und 1 Teil n-Alkanen (*Mepasin*) angeboten, so erhöht sich die Menge der aufgenommenen nichtassimilierbaren Substanz auf das 3—4fache der Menge in Abwesenheit von n-Alkanen. N-Alkane wirken offensichtlich als Schleppersubstanz.

Ergebnisse von Versuchen mit wachsenden Zellen zeigt Tabelle 1.

Tabelle 1. Aromatengehalt wachsender Zellen von *Lodderomyces elongisporus* H in diskontinuierlicher Kultur in Abhängigkeit von der Fermentationszeit

Fermentationszeit (h)	Aromaten		Heterozyklen		Aromaten + Heterozyklen	
	% der HTS	% des Angebotes	% der HTS	% des Angebotes	% der HTS	% des Angebotes
0	0	0	0	0	0	0
2	0,25	0,08	0,12	0,08	0,25	0,05
4	0,17	0,05	0,003	0,002	0,34	0,07
22	0,31	0,09	0,21	0,13	0,24	0,07
25	0,44	0,13	0,56	0,35	0,45	0,09

Eingesetzt wurde ein Modell-Dieselkraftstoff (Gemisch des nichtassimilierbaren Kohlenwasserstoffanteils von DK und *Mepasin*) in Konzentrationen, die jeweils einem Angebot von Original-Dieselkraftstoff von 20 g/l entsprachen.
Bei diskontinuierlicher Kultivierung zeigt sich eine Abhängigkeit des Gehaltes der Zellen an nichtassimilierbaren Kohlenwasserstoffen von der Fermentationszeit.
Die Geschwindigkeit der Aufnahme ist in den einzelnen Wachstumsphasen unterschiedlich. Am höchsten ist sie in den ersten Stunden nach Versuchsbeginn. In den ersten 4 Stunden werden etwa 50% des maximalen Aromatengehalts aufgenommen. In den nächsten 18 Stunden ändert sich der Aromatengehalt nur wenig, erst danach kommt es zu einer weiteren Erhöhung.
Nichtassimilierbare Kohlenwasserstoffe werden auch in wachsende Zellen nur in geringen Mengen aufgenommen, zu einer deutlichen Akkumulation kommt es nicht.
Ein Minimum wird bei der Aufnahme von Aromaten und Heterozyklen offensichtlich während des exponentiellen Wachstums und ein Maximum in der stationären Phase erreicht.

4. *Einfluß von Dimethylsulfoxid (DMSO) auf die Aufnahme von nichtassimilierbaren Kohlenwasserstoffen und n-Alkanen*

DMSO wurde in die Untersuchungen einbezogen im Hinblick auf seine Eigenschaften der Penetrationsförderung und guten physiologischen Verträglichkeit, die auch bei der Applikation von lipophilen Medikamenten genutzt werden [1]. Die Versuche wurden mit *Lodderomyces elongisporus* H durchgeführt. Als Kohlenwasserstoffe wurden das n-Alkangemisch *Mepasin* (Tab. 2) und Dieselkraftstoff-Modelle mit unterschiedlichem n-Alkananteil (Tab. 3) und tritiummarkierter Aromatenfraktion verwendet.

Tabelle 2. Einfluß von DMSO auf die durchschnittliche NFS-Akkumulationsrate in verschiedenen Versuchsstadien $V_A = \frac{(\text{g Akkumulation/g HTS})}{\text{h}}$

Versuchs-stadium	Akkumulationsrate nach Zusatz von DMSO									
	0%		0,01%		0,05%		0,1%		0,5%	
	V_A	% der Kontrolle	V_A	% der Kontrolle	V_A	% der Kontrolle	V_A	% der Kontrolle	V_A	% der Kontrolle
0–4 h	0,14	100	0,41	293	0,55	393	0,36	257	0,51	365
4–12 h	0,14	100	0,13	93	0,05	36	0,09	64	0	0

Tabelle 3. Einfluß von DMSO auf die Aufnahme von nichtassimilierbaren Bestandteilen des Dieselkraftstoffs durch *Lodderomyces elongisporus* H

DMSO-Zusatz in % der Kultur-flüssigkeit	Dieselkraftstoff-Modelle			
	Aromaten 10 g/l + n-Alkane 10 g/l		Aromaten 33,3 g/l + n-Alkane 6 g/l	
	Aromatengehalt		Aromatengehalt	
	% HTS	% Kontrolle	% HTS	% Kontrolle
0	1,37	100	2,95	100
0,005	0,57	42	1,82	62
0,01	1,02	75	2,00	68
0,1	0,95	69	2,02	69
0,5	0,79	53	2,18	74

Auf n-Alkanen bewirkt DMSO in den ersten 4 Stunden des Versuches eine vier- bis sechsfache Beschleunigung der NFS-Akkumulation. Demgegenüber führt auf Dieselkraftstoffmodellen bereits ein Zusatz von 0,005% DMSO zu einer deutlichen Senkung des Aromatengehaltes der HTS im Vergleich zur DMSO-freien Kontrolle (Tab. 3). Die Ursachen für diese völlig unterschiedliche Beeinflussung der Aufnahme von n-Alkanen und Aromaten, d. h. von assimilierbaren und nichtassimilierbaren Kohlenwasserstoffen sind nicht bekannt.

Diskussion

Die Problematik der Kohlenwasserstoffaufnahme beginnt mit der Frage, in welcher Form dieses Substrat an die Zelle gelangen muß, um in einer Geschwindigkeit aufgenommen zu werden, die dem Leistungsvermögen der Zelle entspricht, ob in Tropfen, größer

oder kleiner als die Zelle [2] oder in solubilisierter Form [3]. Es ist wahrscheinlich, daß sich extra- oder perizelluläre Depots aufbauen, aus denen heraus über Solubilisierung die Aufnahme in die Zelle erfolgt. Für beide Schritte — Anlagerung und Feinstverteilung bis Lösung — besitzt oder bildet die zur Kohlenwasserstoffoxidation enzymatisch befähigte Zelle einen Aufnahmeapparat, der nach unseren Beobachtungen nur bei aktiver Stoffwechseltätigkeit voll funktioniert, z. B. nur in Anwesenheit von Sauerstoff und bei optimalen Milieubedingungen für Wachstum oder Produktbildung. Dazu gehören eine spezielle Strukturierung der Zellwand, die eine hohe Adhäsionskraft der Zelloberfläche für Kohlenwasserstoffe bedingt [2, 4, 5, 6] und die Bildung solubilisierender oder emulgierender Substanzen [7, 8, 9, 10]. Gegenstand unserer Arbeit ist die Aufnahme unterschiedlicher Kohlenwasserstoffe, vor allem solcher, die im Gemisch mit nichtassimilierbaren Kohlenwasserstoffen vorliegen.
Dieser Frage ist bisher wenig Aufmerksamkeit geschenkt worden. Munk u. a. [11] gingen davon aus, daß ein Mikroorganismus zwischen utilisierbaren und nichtutilisierbaren Kohlenwasserstoffen nicht unterscheidet und beide Fraktionen mit gleicher Geschwindigkeit und in gleicher Weise aufnimmt. Unsere Ergebnisse widersprechen diesen Vorstellungen, denn wir beobachten eine deutliche Abhängigkeit des Aufnahmeverhaltens sowohl von der Struktur, wie von den Milieubedingungen und den Organismen.
Schon Goma u. a. [3] haben gezeigt, daß n-Alkane unterschiedlicher Struktur unterschiedlich solubilisiert und aufgenommen werden. Gleiche Emulgatoren beeinflussen unterschiedliche Kohlenwasserstoffe in völlig verschiedener Weise [12], was unseren Ergebnissen zum Einfluß von DMSO entspricht.
Hisatsuka u. a., [13] berichten über einen proteinähnlichen Faktor, der möglicherweise Carrierfunktionen ausübt. Die Wechselbeziehungen zwischen Kohlenwasserstoffen, Zellen und Emulgatoren sind so spezifisch, daß die Verhältnisse mit Diffusion oder erleichterter Diffusion unzureichend charakterisiert sind. Finnerty [14] hat darüberhinaus an einem Bakterium gezeigt, daß die Aufnahme energieabhängig ist und gegen ein Konzentrationsgefälle erfolgt, womit wesentliche Kriterien eines aktiven Transports erfüllt sind.
Es mehren sich die Hinweise darauf, daß die Aufnahme von Kohlenwasserstoffen ein von aktiver Stoffwechseltätigkeit abhängiger selektiver Vorgang ist, der über die Milieubedingungen in mannigfacher Weise beeinflußt werden kann.

Eingegangen: 12. 12. 1979

Literatur

[1] Martin, D., Hauthal, H. G.: Dimethylsulfoxid, Akademie-Verlag, Berlin, 1971, S. 445—462.
[2] Miura, Y., Okazaki, M., Hamata, S. I., Murakawa, S. I., Yugen, R.: Biotechn. Bioeng. **19** (1977) 701.
[3] Goma, G., Pareilleux, A., Durand, G.: Agr. Biol. Chem. **38** (1974) 7, 1273.
[4] Käppeli, O., Fiechter, A.: Biotechn. Bioeng. 18 (1976) 967.
[5] Käppeli, O., Fiechter, A.: J. Bact. (1977) 917.
[6] Käppeli, O., Müller, M., Fiechter, A.: J. Bact. **133** (1978) 2, 952.
[7] Goma, G., Al Ani, D., Pareilleux, A.: V. Internat. Ferment. Symp., Berlin 1976
[8] Kulkarni, K., Banett, S. M.: Biotechn. Bioeng. **21** (1979) 337.
[9] Jarvis, F. G., Johnson, M. J.: J. Am. Chem. Soc. **71** (1979) 4124.
[10] Gutierrez, J. R., Erickson, L. E.: Biotechn. Bioeng. **19** (1977) 1331.
[11] Munk, V., Dostalek, M., Volfova, O.: Biotechn. Bioeng. **11** (1969) 383.
[12] Rosenberg, E., Zuckerberg, A., Rubinovitz, C., Gutnick, D. L.: Appl. Environ. Microbiol. **37** (1979) 402.
[13] Hisatsuka, K., Nakahara, T., Minoda, Y., Yamada, K.: Agr. Biol. Chem. **41** (1977) 3, 445.
[14] Finnerty, W. R.: Abstracts XII Intern. Congr. Microbiol. München, 1978.

Acta Biotechnologica 0 (1980) 0, 21–25

The Emulsifying and De-emulsifying Properties of Some Microbial Polysaccharides

D. G. Cooper[1], J. E. Zajic[2] and J. M. Wood[1]

1 Biochemical Engineering
Faculty of Engineering Science
University of Western Ontario
London, Ontario, Canada N6A 5B9

2 Office of the Dean
College of Science
University of Texas at El Paso
El Paso, Texas U.S.A. 79968

Zusammenfassung

Zahlreiche mikrobielle Polysaccharide sind auf ihre Fähigkeiten untersucht worden, Wasser- und Öl-Emulsionen zu brechen oder aber zu stabilisieren.
Negativ geladene schwach verzweigte *o*-Phosphonomannane mit einem relativ hohem Phosphorgehalt zeigen gute de-emulgierende Eigenschaften.
Mannane stellen schlechte De-Emulgatoren dar, aber einige andere neutrale Polysaccharide erreichen fast so gute Werte wie die besten *o*-Phosphonomannane.
Die untersuchten Polysaccharide können nicht als effektive Emulgatoren bezeichnet werden.

Introduction

A few microbial polysaccharides have been shown to have the ability to stabilize emulsions [1, 2, 3]. However, there are no published studies of polysaccharides' ability to break emulsions which have already formed. This paper reports the emulsification and de-emulsification properties of a large number of polysaccharides isolated from either bacteria or yeasts.
The yeasts *Hansenula holstii, H. capsulata, Torulopsis pinus, Pichia pinus* and others produce extracellular *o*-phosphonomannans in good yield [4, 5, 6]. Each yeast produces a distinct type of polymer and there are interspecies variations in the extent of branching, phosphorous content and saccharide linkage. Furthermore, of one of these organisms is grown in a medium which does not contain phosphate it produces a new polymer, usually a neutral mannan, in high yield, instead of the normal phosphonomannan. This group of yeasts provided a series of related polysaccharides for testing for emulsification and de-emulsification abilities.
A variety of neutral bacterial and yeast polysaccharides were also tested.

Materials and Methods

De-emulsification Tests

Standard water and oil emulsions were prepared using synthetic surfactants dissolved in the aqueous phase. For a stable W/O emulsion (water droplets in a continuous oil phase) the aqueous phase contained 0.068% L92 pluronic surfactant. For an O/W (oil in water) emulsion, the aqueous phase contained 0.072% tween 20 and 0.028% span 60. Four ml of either aqueous solution was vortexed with 6 ml of kerosene (oil phase) in a test tube until maximum emulsification was achieved. The polysaccharide to be tested was added (0.01 gm) and the mixture was vortexed for 1 min. The amount of emulsion was measured as a percentage of the total height of the liquids at appropriate time intervals. For each test a plot was made of the natural logarithm of percent emulsion *vs* time and the half-life calculated.

Emulsification Test

The polysaccharide being tested was added (0.01 gm) to 4 ml of distilled water plus 6 ml of kerosene and vortexed in a test tube until maximum emulsification was achieved. The percent emulsion was measured at appropriate time intervals. The tables contain the data for the percentage emulsion after 1 hr and 96 hrs. This indicates the original amount of emulsification and the stability of the emulsion.

Materials

The polysaccharides from organisms *H. holstii* (Y-2448, Y-2154), *H. capsulata* (Y-1842), *T. pinus* (Y-2023), *P. pinus* (Y-2579), *Sporobolomyces sp.* (Y-6493), *Arthrobacter sp.* (B-1459, B-1973) and Y-1401 were from the Northern Regional Research Laboratory, Peoria, II. Kytex L, curdlan and 23/N/O were obtained from Tait and Lyle Ltd. The dextrans were from Pharmacia Fine Chemicals. The gum xanthan was from Sigma Chemical Co. Pullulan was obtained in our laboratory.

Results and Discussions

De-emulsification Properties

The results for the yeast *o*-phosphonomannans and neutral mannans are listed in Table 1. None of the neutral polysaccharides in this group showed any de-emulsification properties. The *o*-phosphonomannans cause varying amounts of emulsion collapse. They were more effective at breaking the water in oil emulsions than the oil in water emulsions. The best were those produced by two strains of *Hansenula holstii*. The *o*-phosphonomannans produced by *Torulopsis pinus* and *Pichia pinus* are both, more highly branched and have lower phosphorous to mannose ratios than the polymer from *H. holstii* [6]. These structural differences could account for the difference in activity. However, the *o*-phosphonomannans from *H. capsulata* has only minimal de-emulsifying properties but its degree of branching and phosphorous to mannose ratio are similar to those for *H. holstii*. There are differences in the saccharide linkages in the two polymers. The *H. holstii* product has mainly 1,2 and 1,3 bonds whereas *H. capsulata* produces 1,2 and 1,6 mannose linkages [6]. Also, the polymer from *H. capsulata* is the only one in Table 1 to have β-1,2 linkages; all the rest are α-1,2 [5]. The acetyl-*o*-phosphonoglucogalactan from *Sporobolomyces* has phosphate groups but these are blocked by acetate functions and therefore not charged. This polymer has no de-emulsification properties which implies that the negative charges on the other *o*-phosphonomannans contributes

Table 1. De-emulsification and Emulsification Properties of o-Phosphonomannans and Related Polysaccharides

	De-emulsification Half-life (hr⁻¹)		Emulsification %	
	w/o	o/w	1 hr	94 hr
Hansenula holstii (Y-2448)				
mannan	>600	>600	16	16
o-phosphonomannan	44	78	9	0
Hansenula holstii (Y-2154)				
o-phosphonomannan	16	386	24	21
Hansenula capsulata (Y-1842)				
mannan	>600	>600	9	10
o-phosphonomannan	287	>600	34	31
Torulopsis pinus (Y-2023)				
mannan	>600	>600	13	7
o-phosphonomannan	227	344	11	8
Pichia pinus (Y-2579)				
glucomannan	>600	>600	8	8
o-phosphonomannan	341	398	16	11
Sporobolomyces sp. (Y-6493)				
acetyl-o-phosphonoglucogalactan	>600	>600	8	8
Control	>600	>600	0	0

w/o — water in oil emulsion (0.068% L92)
o/w — oil in water emulsion (0.072% tween 20 and 0.028% span 60)
hr — hours after vortexing sample

to their effectiveness. Thus the most effective polymers in this group for de-emulsification are charged, have a low degree of branching, a high ratio of phosphorous to mannose and α-1,2 and 1,3 mannose bonds.

Although the mannans did not cause de-emulsification some of the other neutral polysaccharides in Table 2 were more promising. Several of these show good de-emulsification properties although none were as good as the products from *H. holstii*. In general, these polymers, like the o-phosphonomannans, were more effective at breaking the w/o emulsion. For example, pullulan and polysaccharide 23/N/O cause relatively fast coallescence of the w/o emulsion but do not affect the o/w emulsion. However, there are some examples of the opposite trend. Kytex L and polysaccharide Y1401 have no effect on the w/o emulsion but show some ability to break the o/w emulsion.

The four dextrans in Table 2 appear to show a correlation between molecular weight and de-emulsification. The smallest, T10 (approx. molec. wt. 10.000) has no emulsion breaking properties. As the size increases to T500 (approx. molec. wt. 50.000) these polymers become progressively better de-emulsifiers.

Emulsification Properties

In general, the compounds tested were not very good at stabilizing emulsions. The one exception was polysaccharide 23/N/O which was an excellent emulsifier (Table 2). Some of the others, including several of the o-phosphonomannans, caused moderate stabiliza-

Table 2. De-emulsification and Emulsification Properties of Some Neutral Polysaccharides

	De-emulsification Half-line (hr⁻¹)		Emulsification %	
	w/o	o/w	1 hr	94 hr
Leuconostoc mesenteroides				
dextran T10	>600	>600	0	0
dextran T40	159	>600	0	0
dextran T70	113	211	0	0
dextran T500	89	297	0	0
Streptococcus sp.				
curdlan	>600	>600	0	0
Aureobasidium pullulan				
pullulan	46	398	12	8
Xanthamonas campestris				
gum xanthan	534	>600	12	10
Other Polysaccharides				
kytex L	>600	294	0	0
23/N/O	34	>600	100	94
Y-1401	>600	232	9	8
B-1459	198	145	9	9
B-1973	113	>600	29	20
Control	>600	>600	0	0

w/o — water in oil emulsion (0.068% L92)
o/w — oil in water emulsion (0.072% tween 20 and 0.028% span 60)
hr — hours after vortexing sample
B — 1459 from *Arthrobacter* sp.
B — 1973 from *Arthrobacter* sp.

tion. Most of the polysaccharides showed at least minimal activity (5% to 10% emulsion after 94 hr), but some in particular the dextrans, gave no stable emulsion, even after only an hour. Since the emulsification was generally poor, it was not possible to reach any conclusions about correlations between structure and the ability of polysaccharides to stabilize oil and water emulsions.

Eingegangen: 24. 1. 1980

Acknowledgements

This work was supported by the department of Energy, Mines and Resources under the authority of Dr. D. H. SAWATZKY.
We express our appreciation to Dr. M. E. SLODKI for giving us the samples of o-phosphonomannans and mannans.

References

[1] Kaeppeli, O. and Fiechter, A.: J. Bacteriol. **131** (1977) 917.
[2] Rosenberg, E., Zuckerberg, A., Rubinovitz, C. and Gutnick, D. L.: Appl. Environ. Microbiol. **37** (1979) 402.
[3] Rosenberg, E., Perry, A., Gibson, D. T. and Gutnick, D. L.: Appl. Environ. Microbiol. **37** (1979) 409.
[4] Slodki, M. E., Safranski, M. J., Hensley, D. E. and Babcock, G. E.: Appl. Microbial. **19** (1970) 1019.
[5] Slodki, M. E., Ward, R. M., Boundy, J. A. and Cadmus, M. C.: Ferment. Technol. Today (1972) 597.
[6] Seymour, F. R., Slodki, M. E., Plattner, R. D. and Stodola, R. M.: Carbohydrate Research **48** (1976) 225.

Acta Biotechnologica 0 (1980) 0, 27—31

Entwicklung und Toxinbildung von Aspergillus flavus im Festbettreaktor/Füllkörperreaktor

B. Kopp und H.-J. Rehm

Institut für Mikrobiologie, Universität Münster
D-4400 Münster, Tibusstr. 7—15

Summary

The behaviour of the mycelium forming mold *Aspergillus flavus* in packed column culture in comparison to conventional fermentation systems (surface culture and submerged culture) was investigated. The formation of aflatoxin was observed as a marker for the metabolic activity. The results demonstrated that the packed column system is situated with regard to the conditions of cultivation of mycelia forming molds between the surface culture and the submerged culture.

Einleitung

Tauch- und Tropfkörper finden in der Abwasserreinigung Verwendung [1]. Derartige Systeme werden hinsichtlich des bakteriellen Wachstums als Oberflächenkulturen angesehen [2]. Für das Wachstum mycelbildender Pilze, die ständig mit Nährlösung überrieselt werden, liegen keine Untersuchungen vor. Sicherlich müssen andere Bedingungen angenommen werden. Das Ziel der vorliegenden Untersuchungen war es festzustellen, wie sich ein mycelbildender Pilz in Füllkörperkultur im Vergleich zu herkömmlichen Fermentationssystemen (Oberflächen- und Submersverfahren) verhält. Die Aflatoxinbildung wurde als ein Indikator für die Bildung eines sekundären Metaboliten getestet.

Material und Methoden

1. Versuchsanlage

Die Versuchsanlage zeigt die Abb. 1. Sie besteht aus einer senkrechten, mit Gummistopfen verschließbaren Glassäule von 0,75 m Länge und 0,05 m Durchmesser. In dieser Säule wurden als Füllkörper Hochofenschlackensteine so gegeneinander verkantet, daß sie nur etwa die Hälfte des gesamten Säulenvolumens einnahmen. Dadurch hatten die Pilze genügend Raum zur Mycelbildung auf und in dem Material. Die Schlackensteine wurden durch mehrstündiges Auskochen in 0,001 n HCl-Lösung gereinigt und anschließend gewaschen. Dieses Material enthält ausschließlich Oxide, die biologisch nicht verwertet werden [vgl. 3]. Eigene Untersuchungen mit dem Atom-Absorptionsspektralphotometer ergaben, daß nur Spuren dieser Oxide (Mengen geringer als 0,01%) in die Nährlösung abgegeben werden, so daß eine Stoffwechselbeeinflussung auszuschließen ist. Andere Trägermaterialien, wie z. B. Schaumgummi aus Polyurethan, saugfähige

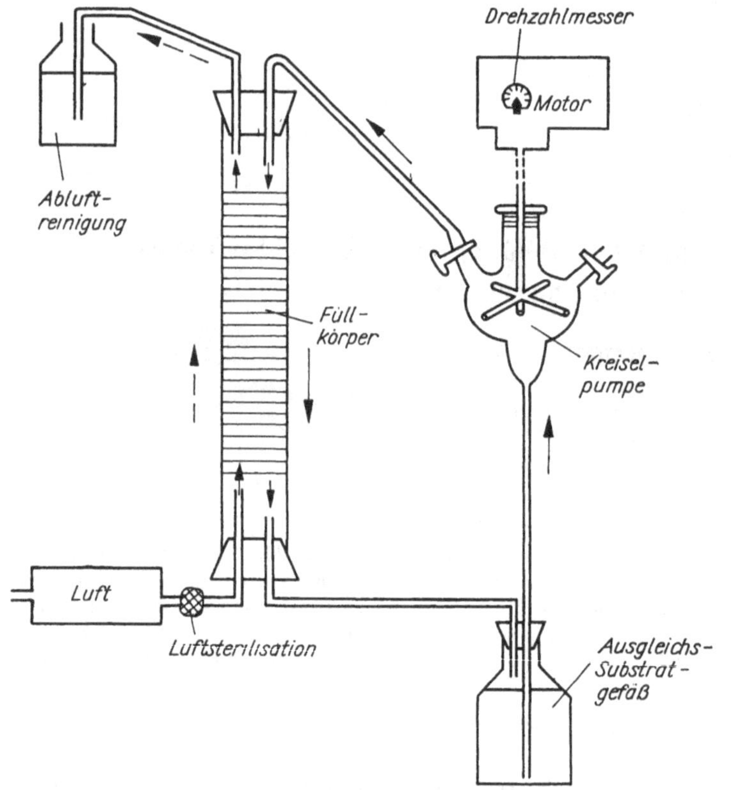

Abb. 1. Versuchsanlage des Füllkörperreaktors.
Weitere Erklärungen im Text

Tonmaterialien hatten sich nicht bewährt. Die Nährlösung rieselte von oben über den mit Mycel bewachsenen Füllkörper und gelangte durch einen Abfluß in das Substratvorratsgefäß. Von hier aus wurde die Flüssigkeit durch eine Kreiselpumpe, die nach dem Zentrifugalsystem arbeitete und durch einen Rührmotor betrieben wurde, angesaugt und im Rücklauf wieder über die Säule gepumpt. Die Durchflußgeschwindigkeit des Nährmediums konnte variiert werden. Für die vorliegenden Untersuchungen wurde in der Regel eine Durchflußgeschwindigkeit von 60 l/h gewählt. Die Belüftung des Systems erfolgte im Gegenstrom. Dabei wurde in der Regel eine Belüftung von 12 l Luft/h gewählt. Die Stoffwechselprodukte wurden aus Proben des Substratvorratsgefäßes bestimmt. Besondere Sicherheitsvorkehrungen hinsichtlich der Arbeiten mit dem Toxin vgl. [4].

2. *Pilzstamm, Inokulum und Nährmedium*

Die Versuche wurden mit *Aspergillus flavus* LINK (CBS 121.62) durchgeführt. Das Inokulum wurde aus 8 Tagen alten Malz-Schrägröhrchen gewonnen; zur Beimpfung von 2 l Medium waren 10^{10} Sporen ausreichend. Als Nährlösung wurde ein komplexes CZAPEK-DOX-Medium [5] eingesetzt. Für eine Füllkörperfermentation wurden 2 l Medium verwendet. Sämtliche Versuche wurden bei 22 °C durchgeführt. Die Oberflächen- und Submerskulturen wurden in 100 ml Medium angezüchtet. Die Submerskultur erfolgte als Schüttelkultur bei 100 Upm.

3. *Aflatoxin-Analytik*

Die Proben der Kulturflüssigkeit wurden jeweils dreimal mit gleichen Volumina Chloroform extrahiert und in 1 ml Chloroform angereichert. Aliquote Mengen dieser Extrakte wurden auf 0,25 mm dicke Kieselgel-Aluminium-Dünnschichtplatten (MERCK) aufgetragen und in dem Fließmittel Chloroform-Aceton (9:1/v:v) entwickelt. Die quantitative Bestimmung der Aflatoxine bei 365 nm konnte direkt auf der Dünnschichtplatte mittels eines Chromatogramm-Spektralphotometers PMQ III (ZEISS) erfolgen [vgl. auch 6, 7].

Ergebnisse

Die Versuchsanordnung stellt ein zweiphasiges, geschlossenes System dar (Rückspeisung der flüssigen Nährlösungsphase von oben, Zufuhr der gasförmigen Phase von unten), in dem sich das Pilzmycel diskontinuierlich auf dem Füllkörper entwickelt. Die Keimhyphen des Pilzes adsorbierten an dem Füllkörper und wuchsen zu einem schleimigen, mehrere Millimeter dicken, die gesamte Oberfläche bedeckenden Mycel aus. Durch den ständigen Überspülungseffekt trat keine Sporulation ein. Die Hochofenschlacke als Füllkörpermaterial ist als biologisch inert anzusehen.

Der mittlere prozentuale Fehler von 5 verschiedenen Fermentationen lag zur Zeit der maximalen Aflatoxinbildung bei 3%. Den Einfluß der Belüftung auf die Stoffwechselaktivität des Pilzes in der Säulenkultur zeigt die Tab. 1.

Die Aflatoxinbildung war weiterhin abhängig von der Durchflußgeschwindigkeit der Nährlösung (vgl. Abb. 2). Die Stoffwechselaktivität des Mycels war bei tropfenweiser

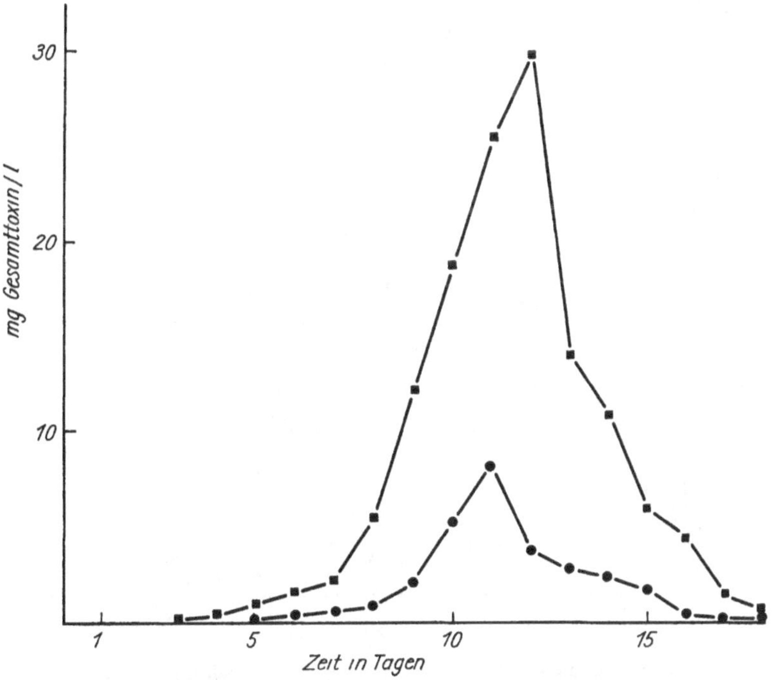

Abb. 2. Abhängigkeit der Aflatoxinbildung von der Durchflußgeschwindigkeit des Substrats (Belüftung 12 l Luft/h)

■ — ■ = Durchflußgeschwindigkeit 0,4 l/h

• — • = Durchflußgeschwindigkeit 60 l/h

Tabelle 1. Maximale Toxinkonzentration in Abhängigkeit von der Belüftungsintensität

Belüftungsstärke	mg Aflatoxin B_1/l	mg Aflatoxin G_1/l.	mg Gesamttoxin/l
0	0,02	0,02	0,04
3,6 l O_2/h	0,20	1,00	1,20
12 l O_2/h	0,70	4,90	5,60

Zufuhr des Mediums (0,4 l/h) wesentlich größer als bei fließendem Mediumsdurchsatz (60 l/h).

Die Abb. 3 zeigt die Unterschiede in den Aflatoxinausbeuten bei der Entwicklung von *Aspergillus flavus* in verschiedenen Fermentationssystemen. Die höchsten Toxinkonzentrationen wurden in Oberflächenkulturen erreicht, die geringsten in Schüttelkulturen, während die Füllkörperkultur eine Mittelstellung einnahm.

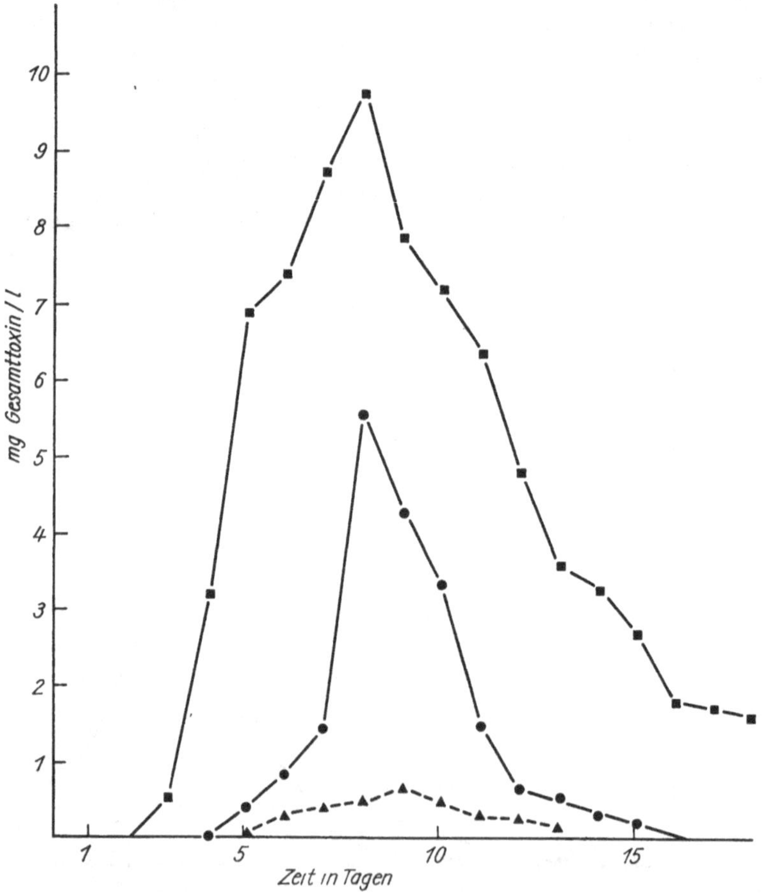

Abb. 3. Unterschiede der Toxinbildung in verschiedenen Fermentationssystemen (Belüftung 12 l Luft/h, Durchflußgeschwindigkeit 60 l/h)

× — × = Oberflächenkultur
• — • = Füllkörperkultur
▲ — ▲ = Submerskultur

Diskussion

Aufgrund der Aflatoxinbildung als „marker" für die Pilzaktivität kann eine Beurteilung des Füllkörperreaktors versucht werden. Wie Abb. 3 zeigt, nimmt hinsichtlich der Toxinbildung die Füllkörperkultur, verglichen mit Oberflächen- und Submersverfahren, eine Mittelstellung ein. Hierfür gibt es verschiedene Erklärungen:

In Submerskulturen kann die Mycelausbildung morphologisch und physiologisch verändert sein [8, 9], so daß es zu einer geringen Aflatoxinsynthese kommt (Abb. 3).

Eine Verminderung der Mycelüberspülungseffekte in Füllkörperreaktoren durch Herabsetzung der Mediums-Durchflußgeschwindigkeit führt zu einer Erhöhung der Toxinrate (vgl. Abb. 2), wodurch eine Annäherung an die Bedingungen einer Oberflächenkultur erreicht ist.

Hinsichtlich der Nährstoffversorgung hingegen kommt das Füllkörpersystem dem optimalen Nährstoffversorgungssystem der Submerskultur nahe, da das Mycel nicht, wie in Oberflächenverfahren, auf ein Nachdiffundieren der Nährstoffe angewiesen ist.

Vom belüftungstechnischen Aspekt her gesehen (vgl. Tab. 1) ähnelt der Füllkörperreaktor jedoch mehr einer Oberflächenkultur, da dem Mycel hauptsächlich Luftsauerstoff angeboten wird und der Gehalt an in dem Medium gelöstem Sauerstoff vergleichsweise gering bleibt.

Das Füllkörpersystem nimmt also hinsichtlich der Kultivierungsbedingungen von mycelbildenden Pilzen eine Mittelstellung zwischen Oberflächen- und Submersverfahren ein. In gewisser Beziehung liegen im Füllkörper bereits trägergebundene Zellen vor.

Eingegangen: 17. 1. 1980

Literatur

[1] Rehm, H. J.: Industrielle Mikrobiologie, 2. Aufl. Berlin—Heidelberg—New York: Springer 1980.
[2] Dietrich, K. R.: Die Abwassertechnik, 2. Aufl. Heidelberg: Dr. Alfred Hüthig 1973.
[3] Römpp, H.: Chemie Lexikon. Stuttgart: Franck'sche Verlagshandlung 1966.
[4] Kopp, B.: Bildung von Aflatoxinen durch *Aspergillus flavus* auf einem Füllkörper. Diplomarbeit Universität Münster 1977.
[5] Wildman, J. D., Stoloff, L., Jacobs, R.: Biotechnol. Bioeng. **9** (1967) 429.
[6] Mann, R., Rehm, H. J.: Europ. J. Appl. Microbiol. **2** (1976) 297.
[7] Mann, R., Rehm, H. J.: Z. Lebensm. Unters. Forsch. **163** (1977) 39.
[8] Trinci, A. P. J.: J. Gen. Microbiol. **67** (1971a) 325.
[9] Trinci, A. P. J.: J. Gen. Microbiol. **67** (1971b) 345.

Acta Biotechnologica 0 (1980) 0, 33–37

On-line Erfassung von Fluoreszenz und Kohlendioxidpartialdruck in Bioreaktoren

A. EINSELE und E. PUHAR

Professur für Mikrobiologie ETH-Hönggerberg Zürich
CH-8093 Zürich

Summary

This paper reports the possibility of on-line-determination of the fluorescence and partial pressure of carbondioxid in bioreactors.
The development of sterilisable probes for contineous in situ measurement of culture fluorescence and pCO_2-value have been reached such a stadium, that both probes are in commercial manufacture.
Fundamental construction of continued developed instruments and the arrangement of probes in bioreactors as known of pH-probes are given; application in each bioreactor with a socket piece of 25 mm diameter is possible.

Einleitung

Die Zuverlässigkeit und Einmaligkeit eines Meßverfahrens, welches in einem Bioreaktor zur Anwendung gelangt, hängt vor allem von der Qualität des Meßfühlers ab. Selbst ein Prozeßrechnersystem ist nur dann optimal eingesetzt, wenn es an zuverlässige Meßfühler angeschlossen ist. Obwohl viele Meß-Sonden theoretisch bekannt sind, sind die Meßeinrichtungen im Routinebetrieb sehr oft auf die Registrierung und allenfalls Regelung von Temperatur, pH und Druck beschränkt. Bereits die zuverlässige Erfassung des gelösten Sauerstoffs im Bioreaktor ist fraglich [5].
Ein Meßverfahren in einem Bioreaktor hat im wesentlichen den folgenden Anforderungen zu genügen:

Die Sonde muß sterilisiert werden können;
damit sind Echtzeitmessungen möglich

Die Sonde muß mit zuverlässiger Genauigkeit geeicht werden können. Im optimalsten Fall sollte das System während des Versuchs (in situ) nachgeeicht werden können.

Die Meßmethodik darf nicht mit anderen Reaktionspartnern interferieren.

Das Instrument darf nicht derart kompliziert sein, daß es nur von speziell geschultem Personal bedient werden kann.

Diese Voraussetzungen wurden konsequent versucht zu erfüllen bei der Entwicklung von Sonden zur in situ-Erfassung der Kulturfluoreszenz und des Kohlendioxidpartialdruckes. Beide Methoden sind so weit entwickelt, daß sie kommerziell erhältlich sind.

Die Erfassung der Fluoreszenz

Die Fluoreszenzmessung einer Nährlösung ist eine Methode, die eine sofortige, direkte, kontinuierliche Erfassung von bestimmten Vorgängen im Bioreaktor erlaubt. Die Kulturfluoreszenz gibt das Oxidations-Reduktionspotential der Zellen an und zwar durch Erfassung des Pyridinnukleotidgleichgewichts (NADH und NADPH). Die ursprüngliche Anwendung der Fluorometrie im Bioreaktor datiert weiter zurück, insbesondere im Zusammenhang mit der Sauerstoffaufnahme von Mikroorganismen.

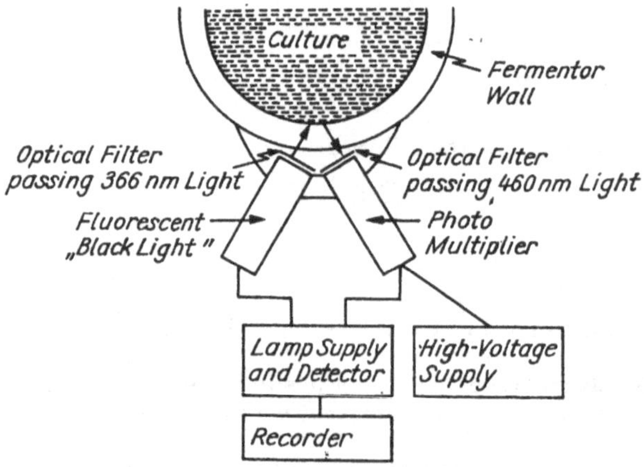

Abb. 1. Prinzip des Fluorimeters im Bioreaktor (vgl. [16])

Prinzipiell stehen durch die in situ-Fluoreszenz-Messung die folgenden Möglichkeiten offen:

a) Bestimmung des Oxidations-Reduktions-Potentials von wachsenden Zellen;
b) Feststellen von Kohlenstofflimitierungen, insbesondere im Zusammenhang mit *Fedbatch*-Züchtungen [12];
c) Bestimmen der aktiven Biomasse [16];
d) Bestimmen der Wachstumsrate von Zellkulturen [16]
e) Bestimmung von Mischzeiten [3];
f) Bestimmung von Substrataufnahmemechanismen [4];
g) Bestimmung von k_La-Werten [14];

Der prinzipielle Aufbau des Instruments ist in Abb. 1 dargestellt. Das Gerät ist normalerweise in einem Schauglas des Bioreaktors angebracht. Die Anordnung sollte zumindest so sein, daß der Lichtstrahl unter der Oberfläche in die Kulturflüssigkeit dringt. Der Rest der Bioreaktorverkleidung muß dunkel sein, damit das übrige Licht die Meßanordnung nicht stört. Die Nährlösung resp. die Mikroorganismen werden mittels einer UV-Lampe, die 366 nm Licht aussendet, bestrahlt. Die so angeregten Stoffe (NADH und NADPH) emittieren Licht von 460 nm, welches über ein Filter- und Sensorsystem aufgefangen und verstärkt wird.
In einer kürzlich erfolgten Weiterentwicklung konnte das optische System miniaturisiert werden (vgl. Abb. 2). Dieses Fluorimeter läßt sich nun ähnlich einer pH-Sonde in den Bioreaktor einbauen. Damit ist diese Meßtechnik für jeden Bioreaktor anwendbar, der über einen Einlaß-Stutzen von 25 mm ⌀ verfügt. Das Licht wird über einen optischen Leiter in den Reaktor geführt. Das dort emittierte Licht wird über einen anderen Licht-

leiter dem Photoverstärker zugeleitet. Das Fluorimeter ist in situ sterilisierbar. Bisherige Versuche haben gezeigt, daß sich damit in allen Reaktoren messen läßt. Vielversprechende Versuche sind im Gange im Hinblick auf die Verwendung des Fluorimeters in *Shift*-Experimenten während Chemostatzüchtungen.

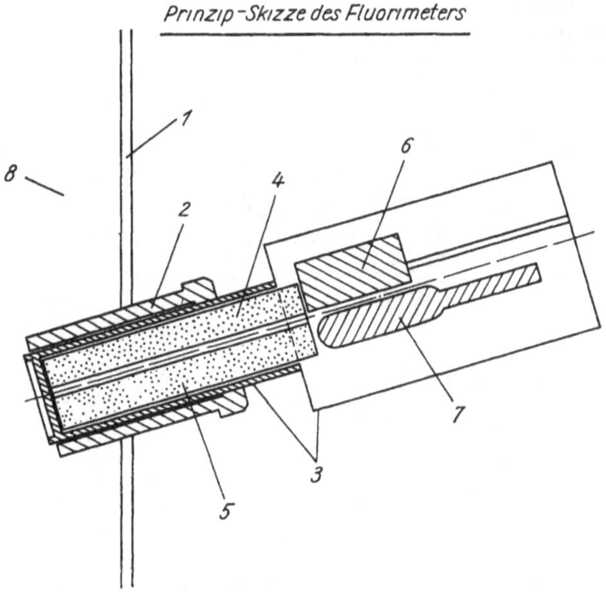

Abb. 2. Prinzip-Skizze des weiterentwickelten Fluorimeters

1: Wand des Bioreaktors
2: Stutzen ∅ 25 mm
3: Umhüllung des Fluorimeters
4: Lichtleiter zur Erfassung der Fluoreszens
5: Lichtleiter zur Projektion des UV-Strahles
6: Optischer Lichtverstärker
7: UV-Lampe
8: Inhalt des Bioreaktors

Die Erfassung des Kohlendioxidpartialdruckes

Obwohl hinlänglich bekannt ist, daß Kohlendioxid einen Einfluß auf Wachstum und Produktbildung und damit auf die Regulation der Zellen hat [1, 2, 7, 8, 11, 13, 17], wird der Kohlendioxidpartialdruck im Bioreaktor selten gemessen. Dies vor allem, weil es an geeigneten Meßmethoden fehlte. Zwei Methoden können zur Messung des pCO_2 in der Flüssigphase verwendet werden:

1) Das CO_2 der Flüssigphase diffundiert durch einen mit Stickstoff-Trägergas durchfluteten Silikonschlauch und wird gaschromatographisch bestimmt.
Diese Methode ist aufwendig [9, 15];

2) Mittels pCO_2-Elektroden (*Severinghaus*-Typen). Diese sind schon seit langem bekannt [6], doch ungeeignete Langzeitstabilität, Nichtsterilisierbarkeit und lange Ansprechzeiten hinderten den Einsatz im Bioreaktor.

In den vergangenen Jahren ist aber die Weiterentwicklung der pCO_2-Sonde sehr stark gefördert worden. Der prinzipielle Aufbau einer pCO_2-Sonde ist in Abb. 3 dargestellt. CO_2 diffundiert durch eine armierte gasdurchlässige Membran, bis auf beiden Seiten gleicher Partialdruck herrscht. Das Gleichgewicht stellt sich sehr rasch ein, wenn eine hochdurchlässige Membran verwendet wird und die Schichtdicke des Bikarbonat-Elektrolyten zwischen Membran und pH-Elektrode klein gehalten wird. Ein Nylonnetz garantiert die konstante Schichtdicke des Elektrolyten.
Verbessertes Membranmaterial sowie spezielle temperaturbeständige pH-Elektroden ermöglichen nun den Einsatz der pCO_2-Elektrode unter sterilen Bedingungen im Bioreaktor [10].

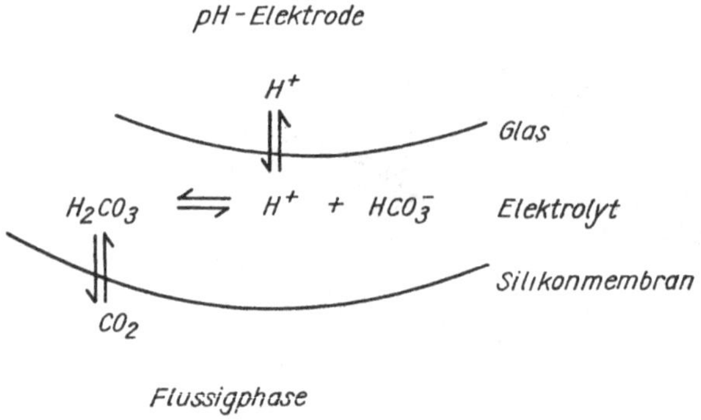

Abb. 3. Aufbau einer pCO_2-Elektrode; Meßgröße ist der pH des Elektrolyten

Ein spezielles Problem stellt die Eichung der pCO_2-Elektrode dar. Diese kann nach zwei Methoden erfolgen:

1) Belüften mit mindestens zwei definierten Gasgemischen und Berechnen der Eichgeraden (mV vs. *log* pCO_2). Dies erfordert schon bei kleinen Reaktoren große Gasmengen und ist während einer Fermentation nicht durchführbar.

2) Die elegantere Methode ist die Beschränkung auf die Überprüfung der pH-Funktion der Glaselektrode, da ein eindeutiger Zusammenhang zwischen der Spannung der pCO_2-Elektrode und derjenigen der Glaselektrode besteht. Dies geschieht mit speziellen Pufferlösungen, die während des Eichvorganges anstelle der Bikarbonatlösung zwischen Membran und Glaselektrode gebracht werden. Die Glaselektrode wird etwas zurückgezogen und der pCO_2-Elektrolyt durch Pufferlösung ersetzt. Die pCO_2-Sonde ist daher so konstruiert, daß die Eichung der Glaselektrode innerhalb der Sonde erfolgen kann, diese somit nicht aus dem Bioreaktor ausgebaut werden muß. Damit stellt eine Nacheichung kein Problem mehr dar.

Langzeitstabilität und Sterilisationsverhalten der pCO_2-Elektrode sind ausgezeichnet [10]. Sie hängen letztendlich nur von der Alterung der pH-Elektrode ab.
Die Elektrode kann in jeden Bioreaktor eingebaut werden, der einen Einlaßnormstutzen von 25 mm ⌀ besitzt.

Eingegangen: 21. 1. 1980

Literatur

[1] BATTAT, E., GOLDBERG, I., MATELES, R. I.: J. of Applied Microbiol. 28 (1977) 906.
[2] CHEN, S. L., GUTMANIS, F.: Biotechnol. Bioengn. 18 (1976) 1455.
[3] EINSELE, A., RISTROPH, D. L. and HUMPHREY, A. E.: Biotechnol. Bioengn. 20 (1978) 1487.

[4] EINSELE, A., RISTROPH, D. L. and HUMPHREY, A. E.: Europ. J. appl. Microbiol. and Biotechnol. **6** (1979) 335.
[5] EINSELE, A.: Swiss Chem. **1** (1979) 19.
[6] FINN., R. K.: ACS-Meeting (1961).
[7] GANDHI, A. P., KJAERGAARD, L.: Biotechnol. Bioengn. **17** (1975) 1109.
[8] NYIRI, L., LENGYEL, Z. L.: Biotechnol. Bioengn. **7** (1965) 343.
[9] PHILLIPS, D. H., JOHNSON, M. J.: J. Biochem. and Microbiol. Technol. and Engn. **3** (1961) 261.
[10] PUHAR, E., EINSELE, A., BÜHLER, H., INGOLD, W.: Biotechnol. Bioengn. (to be published 1980).
[11] REPASKE, R., REPASKE, A. C., MAYER, R. D.: J. Bacteriol. **117** (1974) 652.
[12] RISTROPH, D. L., WATTENEEUW, C. M., ARMINGER, W. B. and HUMPHREY, A. E.: J. Ferment. Technol. **55** (1977) 599.
[13] SHIBAI, H., ISHIZAKI, A., MIZUNO, H., HIROSE, Y.: Agricult. and Biolog. Chem. **37** (1973) 91.
[14] WATTENEEUW, C. M., RISTROPH, D. L. and HUMPHREY, A. E.: ACS-Meeting, Chicago (1977).
[15] YAGI, H., YOSHIDA, F.: Biotechnol. Bioengn. **19** (1977) 801.
[16] ZABRISKIE, D. W. and HUMPHREY, A. E.: Appl. Environ. Microbiol. **35** (1978) 337.
[17] ZADRAZIL, F.: Europ. J. Appl. Microbiol. **1** (1975) 327.

Acta Biotechnologica 0 (1980) 0, 39—46

A study of water-air oxygen equilibrium for the analysis of deep tank aeration

P. MIHÁLTZ and J. HOLLÓ

Technical University, Budapest
Institut of Agricultural Chemical Technology
H-1025 Budapest, Pusztaszeri út 59—67

Zusammenfassung

Die Quellen der zur Berechnung des Sauerstofftransfers herangezogenen Gleichgewichtsdaten des gelösten Sauerstoffs werden kritisch gesichtet und die Grenzen der Verwendbarkeit dieser Daten angegeben. Ein bisher nicht beschriebener Zusammenhang zwischen Temperatur, Gesamtdruck und Sauerstoffpartialdruck wird abgeleitet. Die Triebkraft-Bedingungen der in der biologischen Abwasserreinigung immer breitere Anwendung findenden Tieftank-Belüftung werden graphisch dargestellt, wobei die Zweckmäßigkeit der Anwendung der logarithmischen mittleren Triebkraft aufgezeigt wird. Für stationäre aerobe Systeme wird eine auf den folgenden Erwägungen basierende Methode zur Bestimmung von k_La beschrieben.

Introduction

In the investigation of oxygen transport phenomena in aerobic fermentation processes and the biological treatment of sewage, the formula

$$\frac{dC}{dt} = k_La(C^x - C) - R \tag{1}$$

serves generally as the starting point.

For many years, determination of the overall transfer coefficient characteristic of the rate of aeration has been carried out by a method of saturation in non-steady state. In this case $R = 0$ in equ. (1), i. e., there is no oxygen uptake in the liquid phase. Determination has been carried out by numerical or graphical analysis of $C—t$ data-pairs. In steady state C does not change against time, the left side of the equation is zero. In a biological system R denotes microbiological oxygen requirement. In steady state two terms (oxygen uptake and consumption) on the right side are equal.

In this case the process must be interrupted for the determination of k_La. When aeration is stopped the first term on the right side will be zero and the R value is derived by evaluation of the data-pair $C—t$. When steady state is restored C becomes again and k_La can be calculated from the data on the right side.

Tables of oxygen solubility

As can be seen, exact knowledge of the C^x value is required under all operational conditions or by any method of measurement applied. These values for pure water are generally derived from conventional tables in national standards and are also used for the calibration of membrane oxygen probes.
As regards the data given in the tables as a function of temperature under strictly identical conditions, international usage is far from being uniform.

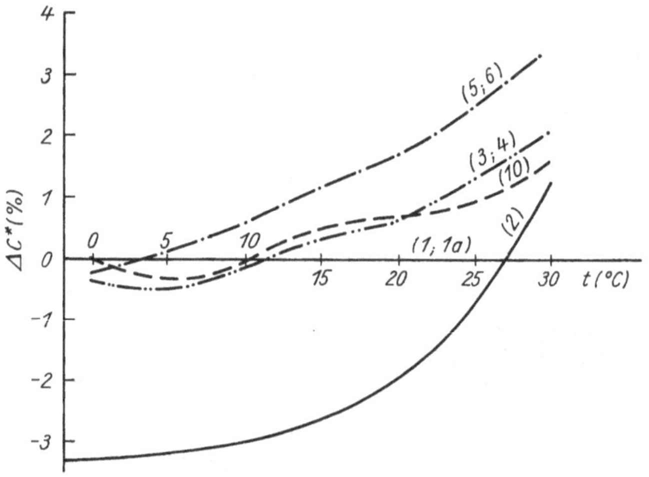

Fig. 1. Deviations between C^x data of different authors

Figure 1 shows the differences (in percentage) between the C^x values given by some authors (manufacturers) or the tables officially used in some industrial countries [2—6, 10] and those applied in CMEA countries on the basis of the CMEA recommenadtion of 1967 [1]. Therefore, in more accurate calculations it is always desirable to indicate the source of the data in order to avoid possible confusion.
In addition, these data give values only for dissolved oxygen in equilibrium under 760 torr total pressure with average oxygen content as given for the troposphere (20, 95 vol.-%), which renders the use of these tables even more restricted.
It is easy to understand why these conditions are untenable, especially in the case of deep tank aeration, a technique widely used in waste water treatment today.
The real value of C^x at the site of air inlet at the bottom of the tanks may be by 50 to 100% higher and on the surface by 20—40% lower than the values depicted in the tables (with 5—10 m water depth and 14—46% oxygen transfer efficiency).

An approximative formula based on Henry's law

Oxygen absorption is a typical field of application for HENRY's law, due to the very poor solubility of oxygen. Values below 1 atm partial pressure (p) of the dissolved gas component are also realized.

As to the use of the simplest form of HENRY's law

$$C^x = \frac{p}{He} \tag{2}$$

or its more general form

$$C^x = \frac{p}{HeRT} \tag{3}$$

the following remarks can be made:

1. In the case of very dilute solutions, the correct expression of C^x as a mole fraction according to HENRY's law is proportional to the corresponding weight fraction without any considerable error. Up to this limit the same holds true for the relation between fugacity and partial oxygen pressure.
2. Owing to the additivity of partial pressures and the proportionality between molar fraction and vol.-%, the partial pressure of oxygen can be described as a product of measurable quantities:

$$p = Pc_t \tag{4}$$

3. Analysis of the data available in the literature show that the temperature dependence of the He constant in the oxygen-water system in the range of 0—30 °C is, in good approximation, linear.
4. Starting from equation (3) we may try to elaborate an approximative formula for one of the conventional tables of equilibrium dissolved oxygen, with the use of the implicit fixed partial oxygen pressure (0.2095 atm).

It is not necessary to give the dimensions of constant He and the universal gas constant. For numerical evaluation of the $C_1^xT_1$ values of the CMEA table [1], we studied the dependence of the product (HeR) on the absolute temperature. With the use of equ. (3), our above point 2., partial pressure N/m², absolute temperature K° and dimensionless $C^x{}_1$ weight fraction:

$$(HeR) = \frac{Pc_t}{C_1^xT_1} = \frac{20\,950}{C_1^xT_1} \tag{5}$$

in the temperature range 0—30 °C (273—303 °K) the data can be fitted to the equation of the straight line

$$(HeR) = 0.13534 \cdot 10^6\, T_1 - 31.729 \cdot 10^6 \tag{6}$$

by a correlation factor of 0.999984. At 99% significance level and $\nu = 16$, the value of the slope is within the reliability range of $\pm 0.043\%$ (Student's statistical test).
Then expression (6) can be substituted into formula (3) to obtain an approximative relationship which takes into account also the effect of P, c_t and temperature:

$$C^x = \frac{Pc_t}{0.1353 \cdot 10^6\, T^2 - 31.73 \cdot 10^6\, T} \tag{7}$$

Comparison of values C_1^x calculated by relationship (7) and those given in Table [1] gives an error below 0,58% in the temperature range 0—30 °C.

An example for deep tank oxygenation with activated sludge

The above relationship shows the practical role of factors affecting the C^x value. Our aim has been to establish accurately by means of equ. (7) the driving force conditions determining the rate of oxygen transfer. The driving force expressed in equ. (1) simply as the difference $(C^x - C)$ may, namely, lead to erroneous results.

To present an example, we examined the oxygen transport processes of two activated sludge tanks of typical depths aerated by bubbling.

In the presence of dissolved and suspended contaminants, the known β factor, which expresses the deviation from the pure water-air system, must also be taken into consideration (7):

$$C_s^x = \beta \frac{Pc_t}{0.1353 \cdot T^2 - 31.73T} \tag{8}$$

where C_s^x is expressed in terms of g/m³ or in mg/l units. Let us presume oxygen transfer efficiency to be $0_{1-2} = 23\%$ and $0_{11-12} = 41\%$ in tanks of 5 and 10 m depth, respectively. Thus, the oxygen content of gas bubbles leaving the liquid phase is stabilized as $c_{t2} = 0.17$ and $c_{t12} = 0.135$ v/v, respectively. The driving force conditions of oxygen transfer are presented in a diagram containing also the above symbols where the oxygen contents of gas and liquid phase are indicated on the axes (Fig. 2). Equilibrium compositions are plotted by oblique straight lines starting from the origo, indicating as parameters total pressures corresponding to various water depths of $y = 0-10$ m at a given temperature (18 °C; $\beta = 0.956$). The approximation accepted for bubbled tanks is the socalled plug-flow behaviour of the gas phase; thus, statistically, one particular bubble loses oxygen on its ascending course to an extent determined by the rate of transfer. The behaviour of the liquid phase is accepted, on the other hand, as a completely mixed state [8, 9]. Accordingly, in steady state microbe respiration may develop an identical level of dissolved oxygen at every point of the tank, in our example the typical value: $C_1 = C_2 = C_{12} = 2$ mg/l. Naturally, in addition to hydrodynamical conditions, this requires also uniform distribution of sewage inlet and outlet in the tank. Points 1; 2 and 11; 12 characteristic of the bottom and surface of 5 and 10 m tanks

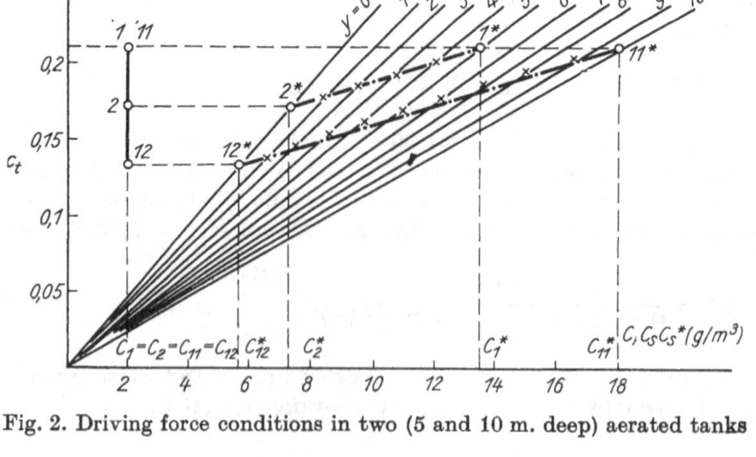

Fig. 2. Driving force conditions in two (5 and 10 m. deep) aerated tanks

can be indicated in the diagram by the given data of phase composition. The lines connecting these points may be denoted as operating lines characteristic of changes in the composition of the gas phase and the liquid environment (1—2 and 11—12, respectively).

It is easy to see that the cross section of a given height of the basin corresponds to any point of these lines. From the equilibrium lines pertaining to corresponding water depths, the horizontal lines drawn from the end points of the operating lines cut out, under given conditions, the equilibrium composition of the liquid phase. The lines $1^x—2^x$ and $11^x—12^x$ obtained in this way are denoted as equilibrium operating lines. The local driving force of transfer is equal to the horizontal distance between the point pairs of the operating line and equilibrium operating line. It can be seen that in both cases the driving force undergoes considerable changes between the surface and the bottom. Typical values of water depths and oxygen transfer efficiency are based on reliable experimental data [8, 9]. It was also found that the oxygen content in the gas phase showed an approximately linear relationship with the length of the path made by the bubble. On the basis of this, we established the real points of equilibrium operating lines on consecutive equilibrium lines between end points $1^x—2^x$ and $11^x—12^x$ in identical vertical steps of Δc_t · As can be seen, even at 10 m water depth there was only slight deviation from the straight line drawn by points and dashes.

Between two straight lines, on the other hand, the known logarithmic formula can be used as a correct expression of the average driving force, e.g., in the first case:

$$\Delta C_{1-2}\ \lg = \frac{C_1^x - C_1 - (C_2^x - C_2)}{\ln \dfrac{C_1^x - C_1}{C_2^x - C_2}} \tag{9}$$

In case of a perfectly mixed liquid phase, this can be further simplified, since $C_1 = C_2$:

$$\Delta C_{1-2}\ \lg = \frac{C_1^x - C_2^x}{\ln \dfrac{C_1^x - C_1}{C_2^x - C_2}} \tag{10}$$

The fulfilment of latter condition was considered as proven experimentally by Jackson and Shen [9]. In the experimental part of their communication, devoted to the evaluation of measurements, they compare formulas derived analytically and determined by various approximations with the simplest expression most generally applied for the determination of the arithmetical mean equilibrium of dissolved oxygen:

$$\bar{C}^x = \frac{P_1 c_{t1} + P_2 c_{t2}}{2H} = \frac{C_1^x - C_2^x}{2} \tag{11}$$

In a biological system equ. (10) is suitable above equilibrium for the expression of the real average driving force (the case of $C > 0$).

From the convincing experimental data of Jackson and Shen the following simple approximation can be derived for the depth dependence of the oxygen content of the ascending bubbles

$$c_t = 0.2095 - 0.0051y \tag{12}$$

With the use of formula (12), Fig 3 plotted on the basis of the data of Fig 2, shows the deviation between the mean driving force expressed by equ. (10) and the arithmetical mean ($c^x - C$) expressed by formula (11).

Considerable deviation can be observed even in case $C = 0$, i.e., in sulphite oxidation measurement at considerable tank depths.
Jackson et al. [11] also studied the effect of a porous air diffuser generating essentially greater interfaces than obtained in the cited communication [9]. With this ceramic diffuser substantially higher O values were observed at identical lengths of bubble path. It may be presumed, therefore, that the approximation according to (12) is valid also for higher C values charakterizing the steady state of the biological system. In this case, however, the deviations indicated in the figure show an extremely high increase, which renders the use of the arithmetical mean driving force irrealistic.

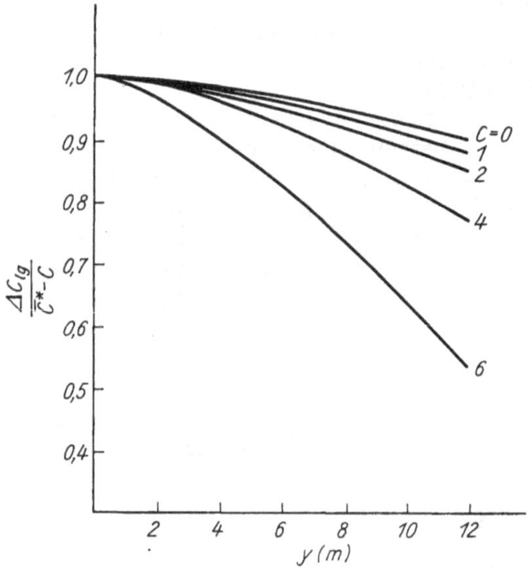

Fig. 3. Ratio of logarithmical and arithmetical driving forces at various water depths and dissolved oxygen levels.

Conclusions

Taking into account steady state caused by microbe respiration, and our above considerations, we can use equ. (1) as follows:
The left side is equal to zero, whereas the two terms on the right are equal to eachother:

$$R = k_L a (C^x - C) \tag{13}$$

Instead of the term in brackets the real logarithmical driving force must be applied:

$$R = k_L a \frac{C^x_{be} - C^x_{ki}}{\ln \dfrac{C^x_{be} - C_{be}}{C^x_{ki} - C_{ki}}} \tag{14}$$

The value of R can be obtained from the oxygen balance of the air introduced into the system:

$$R = \frac{Q(c_{tbe} - c_{tki})\, \varrho}{(1 - c_{tki})\, V} \tag{15}$$

whereas the value of $k_L a$ can be derived from equations (14) and (15):

$$k_L a = \frac{Q(c_{tbe} - c_{tki})\,\varrho \ln \dfrac{C^x_{be} - C_{be}}{C^x_{ki} - C_{ki}}}{V(C^x_{be} - C^x_{ki})\,(1 - c_{tki})} \tag{16}$$

In a practical case, the determination of $k_L a$ requires the measurement of dissolved oxygen in the liquid — and the oxygen content of the gas phase leaving the system, which can be carried out by means of an electrode measuring the partial pressure of oxygen. In addition to this, only the amount of air flow (Q) into the system has to be measured.

If required, e.g., in a sewage sample disinfected and saturated with air, the value of β can be determined by means of the WINKLER method. If, due to unsatisfactory mixing, measurements in the liquid phase show some slight difference between the oxygen content at the bottom and the surface of the tank, the formula of the ligarithmic mean driving force can be substituted by the generalized form given in equ. (9).

Obviously, calculations can be carried out without plotting Fig. 2., used here for the presentation of oxygen transfer conditions. In this case, expression (8) is substituted for the values C^x_{be}; C^x_{ki} and an explicit formula of evaluation is obtained which is suitable for routine tests and can be readily solved by means of a programmable pocket calculator.

Symbols

C; C_s level of dissolved oxygen (in sewage) dimensionless weight fraction or g/m³

C^x; C_s^x level of dissolved oxygen (in sewage) in equilibrium with the gas phase

c_t oxygen content of gas phase (v/v)

He HENRY's constant

$k_L a$ overall oxygen transfer coefficient (1/hr)

o oxygen transfer efficiency (%) $\left(\dfrac{c_{t1} - c_{t2}}{c_{t1}(1 - c_{t2})} \cdot 100\right)$

p partial pressure of oxygen (N/m²)

P total pressure (N/m²)

Q air flow (Nm³/hr)

R rate of oxygen uptake (g/m³ hr); universal gas constant

T absolute temperature, °K

t time (hr)

y vertical dimension, liquid depth (m)

V volume of water in the basin (m³)

$\beta = \dfrac{C_s^x}{C^x}$ factor expressing the decrease in equilibrium dissolved oxygen

ϱ density of gaseous oxygen (g/m³)

ΔC_{lg} average logarithmical driving force (g/m³)

Indexes (subcripts)

be; 1; 11 oxygen input at the bottom of tank

ki; 2; 12 oxygen output at the surface of tank

Eingegangen: 5. 12. 1979

References

[1] Standard methods of water analysis, COMECON, Vol. 1; VITUKI (Scientific Research Institute for Water Economy) Budapest, 1968.
[1a] Radelkis: Aquacheck instrument book.
[2] Truesdale, G. A., Downing, A. L., Lowden, G. F.: J. Appl. Chem. 5 (1955) 53.
[3] Mortimer, C. H.: Comm. No. 6. Ass. Int. de Limnologie Théorique et Appliquée, Stuttgart (1956).
[4] Norme Francaise Enrigistrée T 90-032 (avril, 1975).
[5] Standard Methods for the Examination of Water and Wastewater. Amer. Publ. Health Ass. Inc. New York (1965).
[6] Beckmann Co.: Fieldlab Dissolved Oxygen Analyzer (Instrument Book).
[7] Van der Kroon: Proc. 4th Internat. Conf. on Water Pollution Research, Prague, 2–6, 9., 1968.
[8] Urza, I. J., Jackson, M. L.: Ind. Eng. Chem. Process Des. Dev. V. 14 No. 2 (1975) 108.
[9] Jackson, M. L., Shen, C. C.: AiChE Journal, V. 24., No. 1. (1978) 63.
[10] Knop, E., Bischofsberger, W., Stahlmann, V.: Versuche mit verschiedenen Belüftungssystemen in technischem Maßstab. Teil I. Vulkan Verlag, Essen (1964).
[11] Jackson, M. L. et al.: AIChE Symp. Series Water — 1975 159. No. 151, 71.

Acta Biotechnologica 0 (1980) 0, 47–54

Einfluß von Zustandsvariablen auf die Erhaltungs- und Verbrauchskoeffizienten bei mikrobiellen Prozessen

J. D. Schneider

Akademie der Wissenschaften der DDR
Institut für technische Chemie Leipzig
7050 Leipzig, Permoserstr. 15

Summary

It is shown for different examples of microorganisms growth in continuous flow culture the influence of process variables (temperature, H^+-concentration and changes of the chemical composition of fermentation medium) on the coefficients of maintenance and substrate consumption.
The maintenance coefficient is increased exponentially with increasing temperature or H^+-concentration.
The minimum maintenance coefficient is obtained when the growth rate is limited by carbon sources.
Other growth limitations or change of chemical composition in the medium lead to rising of maintenance coefficient but not in each case to increasing of specific C-substrate consumption at lower residence time, especially.

1. Einleitung

Kontinuierlich geführte Fermentationen zur aeroben Vermehrung von Mikroorganismen sind unter chemostatischen Bedingungen und bei genetischer Stabilität der Mikroorganismen durch ein Fließgleichgewicht charakterisiert.
Ausdruck solcher Fließgleichgewichte ist die energetische und stoffliche Effizienz der mikrobiologischen Stoffwandlung. Der Wirkungsgrad oder Ausbeutekoeffizient solcher irreversibler Reaktionen bei einem fixierten Mikroorganismen/C-Substrat-System hängt in entscheidendem Maße von den Zustandsvariablen bzw. äußeren Zustandsbedingungen ab. Die Zustandsvariablen wie Temperatur und die chemische Zusammensetzung (Konzentrationen an essentiellen chemischen Komponenten sowie die H^+-Konzentration) bestimmen die Wachstumsrate und die Ausbeutekoeffizienten. Letztere sind durch die allgemein gültige Beziehung

$$Y^{-1} - Y_0^{-1} = m \cdot \mu^{-1} \quad (1)$$

verknüpft [1].
Jede differentielle Änderung einer Zustandsvariablen führt zu einem neuen Fließgleichgewicht und umgekehrt induzieren Änderungen der Wachstumsrate neue äußere Zustandsbedingungen (z. B. extrazelluläre Konzentrationen sowie veränderte Utilisationsgeschwindigkeiten der Reaktionskomponenten).

Nach dem Stationäritätsprinzip des Chemostaten wird die Wachstumsrate durch die Verdünnungsrate D festgelegt. Es gilt:

$$\mu = D \quad \text{bzw.} \quad D = 1/t_R$$

Anstelle des Ausbeutekoeffizienten Y wird häufig der Verbrauchskoeffizient $\alpha_x{}^S$ benutzt ($\alpha_x{}^S = 1/Y_S$). Unter Verwendung der spezifischen Größen $\alpha_x{}^S$ und t_R folgt aus Gleichung 1 die allgemeine Zustandsfunktion:

$$\alpha_x{}^S = \alpha_{x_0}^S + m_S \cdot t_R \tag{2}$$

Bei Kenntnis der zwei Größen $\alpha_{x_0}^S$ und m_S kann für beliebige mittlere Verweilzeiten der Verbrauchskoeffizient eines Substrates berechnet werden.
Der Erhaltungskoeffizient m_S für das C-Substrat ist stets größer als Null ($m_S > 0$), die bei $t_R = 0$ extrapolierte Größe $\alpha_{x_0}^S$ ist in der Regel positiv, kann aber auch bei nicht-optimalen Prozeßbedingungen negative Werte annehmen.
Im folgenden wird der Einfluß der Temperatur, des pH-Wertes und einiger chemischer Komponenten auf den Erhaltungs- und Verbrauchskoeffizienten an Beispielen dargestellt.

2. Temperaturabhängigkeit des Verbrauchs- und Erhaltungskoeffizienten

Nach Untersuchungen von TOPIWALA und SINCLAIR [2] ist der Erhaltungskoeffizient von der Temperatur nach einer e-Funktion abhängig:

$$m_S = K_1 \cdot e^{-\Delta H/RT} \tag{3}$$

Nach Kombination der Gleichungen 2 und 3 erhält man den Zusammenhang zwischen den Verbrauchskoeffizienten und der absoluten Temperatur:

$$\alpha_x{}^S = \alpha_{x_0}^S + K_1 \cdot e^{-\Delta H/RT} \cdot t_R \tag{4}$$

Wird die Fermentation bei einer bestimmten Verweilzeit ($t_R = \text{const.}$) durchgeführt, dann kann das Produkt aus $K_1 \cdot t_R$ zur allgemeinen Konstante K zusammengefaßt werden.
Nach Logarithmieren von Gleichung 4 wird die lineare Gleichung 5 erhalten:

$$\log (\alpha_x{}^S - \alpha_{x_0}^S) = \log K - \frac{\Delta H}{2{,}3 \cdot R \cdot T} \tag{5}$$

Experimentelle Ergebnisse zeigen eine gute Übereinstimmung mit Gleichung 5 im superoptimalen Temperaturbereich ($T_i > T_{\text{opt.}}$). (vgl. Abb. 1).

Tabelle 1. Reaktionsenthalpien verschiedener Mikroorganismen-Substrat-Systeme im superoptimalen Temperaturbereich

Mikroorganismus	C-Substrat	ΔH (kcal/mol)
Gluconobacter spec.	Methanol	77
Lodderomyces elongosporus	Erdöldestillat	46
Bakterien spec.	Erdöldestillat	36
Hansenula polymorpha	Methanol	44

(Daten nach COONEY et al. [3])

Die berechneten Reaktionsenthalpien der vier untersuchten Mikroorganismen-Substrat-Systeme liegen zwischen 30 und 80 kcal/mol und sind deutlich höher als die entsprechenden Reaktionsenthalpien für den suboptimalen Temperaturbereich ($T_i < T_{opt.}$) ($\Delta H = -15$ bis -20 kcal/mol).
Die Daten der Tabelle 1 zeigen, daß die Reaktionsenthalpien von der Art der Mikroorganismen und weniger vom benutzten C-Substrat abhängig sind.

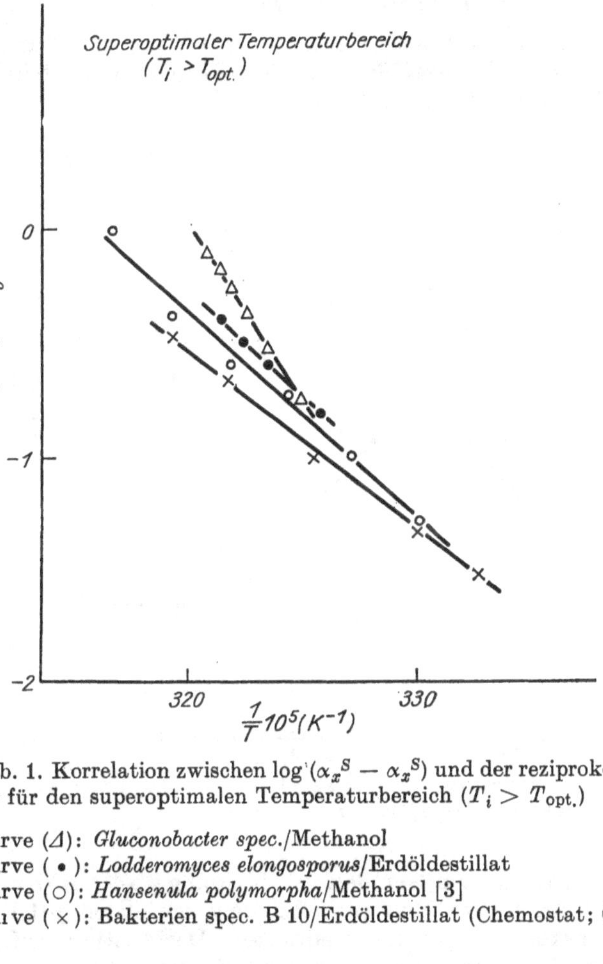

Abb. 1. Korrelation zwischen log ($\alpha_x^S - \alpha_x^S$) und der reziproken absoluten Temperatur für den superoptimalen Temperaturbereich ($T_i > T_{opt.}$)

Kurve (Δ): *Gluconobacter spec.*/Methanol
Kurve (•): *Lodderomyces elongosporus*/Erdöldestillat
Kurve (○): *Hansenula polymorpha*/Methanol [3]
Kuıve (×): Bakterien spec. B 10/Erdöldestillat (Chemostat; C-Limitation)

3. pH-Abhängigkeit des Verbrauchs- und Erhaltungskoeffizienten

Änderungen der Wasserstoffionenkonzentrationen im extrazellulären Bereich gegenüber der optimalen H^+-Konzentration zeigen mehr oder weniger signifikante Erhöhungen der Verbrauchskoeffizienten. Im allgemeinen reagieren Bakterien wesentlich empfindlicher auf pH-Änderungen als Hefen, die in der Regel über einen ausgedehnten pH-Bereich nahezu konstante Verbrauchskoeffizienten besitzen.
Zur Beschreibung der zwei funktionellen Bereiche um den optimalen pH-Punkt wird angenommen, daß der Erhaltungskoeffizient exponentiell von der extrazellulären

H^+-Konzentration abhängig ist:

$$m_S = K_2(C_{H^+})^{K_3} \tag{6}$$

Die Kopplung der Gleichungen 6 und 2 führen zur allgemeinen Beziehung:

$$\alpha_x{}^S = \alpha_{x_0}^S + K_2 \cdot (C_{H^+})^{K_3} \cdot t_R \tag{7}$$

Gleichung 7 logarithmiert ergibt:

$$\log\,(\alpha_x{}^S - \alpha_{x_0}^S) = \log K_2 \cdot t_R - K_3 \cdot pH \tag{8}$$

Die Korrelation zwischen $\log\,(\alpha_x{}^S - \alpha_{x_0}^S)$ und dem pH-Wert weist für die zwei untersuchten Mikroorganismen-C-Substrat-Systeme jeweils ein symmetrisches Verhalten um den optimalen pH-Wert auf (vgl. Abb. 2).

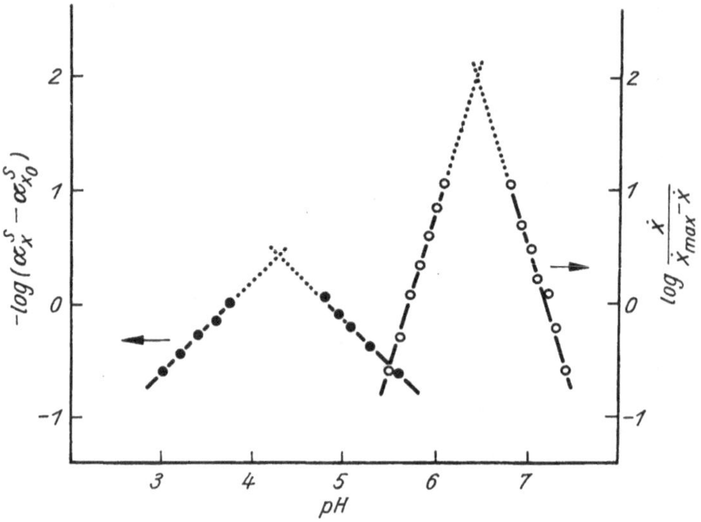

Abb. 2. Korrelation zwischen $-\log\,(\alpha_x{}^S - \alpha_{x_0}^S)$ bzw. $\log \dfrac{\dot{x}}{\dot{x}_{\max} - \dot{x}}$ und dem pH-Wert ($\alpha_x{}^S = \dfrac{S_E}{t_R \cdot \dot{x}}$ bzw. $\alpha_{x_0}^S = \dfrac{S_E}{t_R \cdot \dot{x}_{\max}}$ bei S_E, $\dot{x}_{\max}$, t_R = const.)
Kurve (•): *Gluconobacter spec.*/Methanol
Kurve (○): Bakterien GB 4/Methan

Der Schnittpunkt der verlängerten Geraden in Abbildung 2 ist mit dem experimentell ermittelten optimalen pH-Punkt identisch. Bei dem optimalen pH-Punkt sind die spezifischen Verbrauchskoeffizienten am niedrigsten. Der Anstiegskoeffizient K_3 charakterisiert die pH-Empfindlichkeit eines Mikroorganismus.

4. Einfluß der chemischen Zusammensetzung auf den Verbrauchs- und Erhaltungskoeffizienten

Die Ökonomie der mikrobiellen Zellsubstanzsynthese (SCP-Produktion) wird wesentlich durch den Preis der zugesetzten C-Quelle und durch den spezifischen C-Substratverbrauch ($\alpha_x{}^S$) bestimmt. Gemäß Gleichung 2 werden diese Prozesse vorteilhaft bei möglichst hohen Verdünnungsraten bzw. bei kleineren Verweilzeiten durchgeführt.

Minimale Verbrauchskoeffizienten werden demnach in der Nähe von D_{crit} bzw. $t_{R\min}$ erhalten.
Der Erhaltungskoeffizient m_S ist minimal, wenn die Wachstumsrate der Mikroorganismen im Chemostaten durch die C- und Energiequelle begrenzt wird.
Verschiedene Untersuchungen zeigten jedoch, daß minimale Verbrauchskoeffizienten $\alpha_x{}^S$ nicht in jedem Falle unter den Bedingungen, daß t_R und m_S minimal sind, erreicht werden. Die Verbrauchskoeffizienten können durchaus noch erniedrigt werden, sofern die Extrapolationsgröße $\alpha_{x_0}^S$ unter bestimmten Zustandsbedingungen stärker erniedrigt wird als das Produkt aus $m_S \cdot t_R$.
Dieses interessante Verhalten verschiedener Mikroorganismen-Substrat-Systeme soll an drei Beispielen näher erläutert werden:

- C- und O_2-Limitation bei der Methanassimilation
- Einfluß einer oberflächenaktiven Substanz (*Tensid*)
- Einfluß eines zweiten C-Substrates.

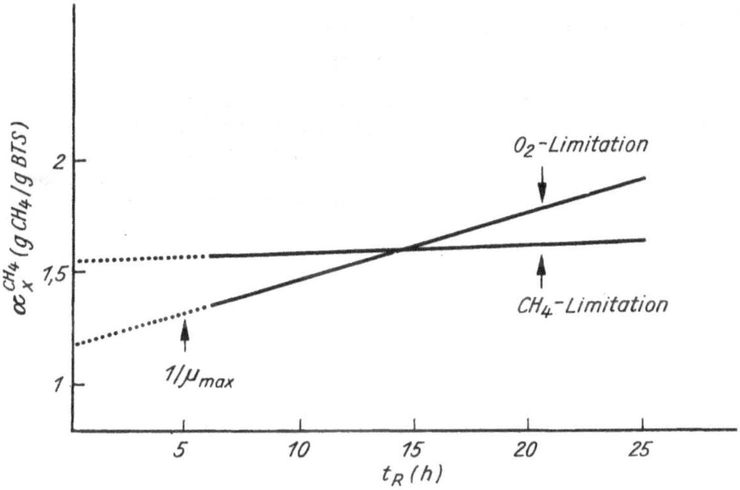

Abb. 3. Abhängigkeit des CH_4-Verbrauchskoeffizienten von der Verweilzeit unter den Bedingungen einer O_2- und CH_4-Limitation (Bakterienkultur GB 21)

C- und O_2-Limitation am Beispiel der Methanassimilation mit der Bakterienkultur GB 21

Unter den Bedingungen einer CH_4-Limitation im Chemostaten betragen die Koeffizienten:

$$m_S = 0{,}007 \text{ g } CH_4/\text{g } BTS \cdot \text{h}$$

$$\alpha_{x_0}^S = 1{,}53 \text{ g } CH_4/\text{g } BTS$$

Der Erhaltungskoeffizient ist gegenüber anderen Limitationsarten minimal.
Bei O_2-limitiertem Chemostatenprozeß erhöht sich der Erhaltungskoeffizient um etwa das 5-fache, der $\alpha_{x_0}^S$-Wert wird jedoch deutlich erniedrigt:

$$m_S = 0{,}034 \text{ g } CH_4/\text{g } BTS \cdot \text{h}$$

$$\alpha_{x_0}^S = 1{,}15 \text{ g } CH_4/\text{g } BTS$$

In Abbildung 3 sind die funktionellen Zusammenhänge nach Gleichung 2 unter den Bedingungen einer C- und O_2-Limitation grafisch veranschaulicht. Die Kurven schneiden sich bei $t_R = 14{,}1$ h.

Das heißt bei $t_R < 14$ h ist der Chemostatenprozeß unter O_2-Limitation vorteilhafter als unter CH_4-Limitation, die trotz geringerem Erhaltungskoeffizienten erst oberhalb $t_R > 14$ h effektiver als die O_2-Limitation wird.

Einfluß eines speziellen Tensides

Es ist gut bekannt, daß spezielle längerkettige nichtionogene oberflächenaktive Substanzen (*Tenside*) die Sauerstoffübergangsrate Gas/Flüssigkeit erhöhen können [4, 5].
Uns interessierte die Frage, ob und in welchem Maße *Tenside* den Erhaltungskoeffizient beeinflussen können.
Ein Zusatz der oberflächenaktiven Substanz WOFK (Polyethylenoxidaddukt des Kokosfettalkohols) führt zur deutlichen Erhöhung des Erhaltungskoeffizienten m_S bei gleichzeitiger Abnahme der Extrapolationsgröße $\alpha^s_{x_0}$ (System: Ethanol/*Candida utilis*; C-Limitation).
Die ermittelten Koeffizienten betragen:

ohne WOFK: $\alpha^s_{x_0} = 1{,}56$ g Ethanol/g HTS

$m_S = 0{,}01$ g Ethanol/g $HTS \cdot$ h

mit WOFK: $\alpha^s_{x_0} = 1{,}24$ g Ethanol/g HTS

$m_S = 0{,}056$ g Ethanol/g $HTS \cdot$ h

Nach Gleichung 2 resultieren aus diesen Daten zwei Geraden, die sich im untersuchten Verweilzeitbereich von $t_R = 2{,}5$ bis 12 Std. bei $t_R = 7$ Std. schneiden (vgl. Abb. 4).
Obwohl der Zusatz der chemischen Komponenten WOFK den Erhaltungskoeffizient um das ca. 5fache erhöht, werden im Verweilzeitbereich zwischen 2,5 und 7 Std. geringere Ethanol-Verbrauchskoeffizienten erhalten als bei dem tensidfreien System.

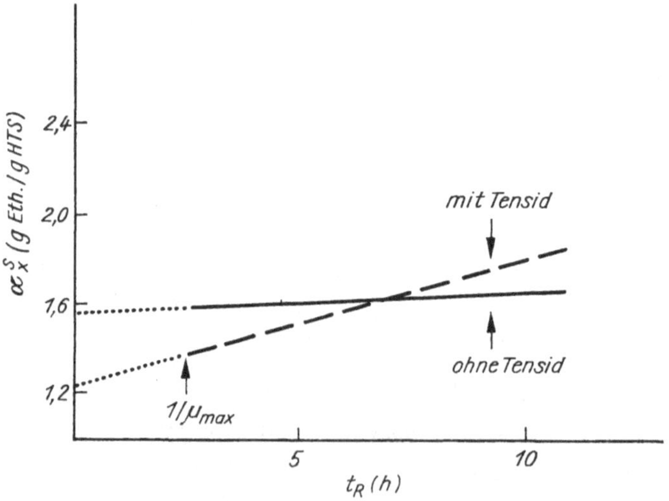

Abb. 4. Abhängigkeit des Ethanol-Verbrauchskoeffizienten von der Verweilzeit mit und ohne Zusatz des Tensides WOFK (System: *Candida utilis*/Ethanol; C-Limitation)

Essigsäurezusatz bei der Saccharosefermentation

Eine Reihe von Mikroorganismen sind fähig, sowohl Saccharose bzw. Glukose als auch Essigsäure als Einzelkomponenten zu assimilieren [6, 7]. Bei den technischen Verfahren der Sulfitablaugeverhefung ist gut bekannt, daß Kohlenhydrate und Essigsäure parallel gut utilisiert werden können.

Aus der Literatur geht hervor, daß die Erhaltungs- und Verbrauchskoeffizienten der Essigsäureassimilation stets höher als die entsprechenden Koeffizienten der Saccharoseassimilationen sind. Es liegen bisher keine Ergebnisse vor, in welchem Maße die Koeffizienten der Essigsäure-Saccharose-Gemische sich von denen der Einzelkomponenten unterscheiden.
Die Ergebnisse der Verhefung von reiner Saccharose, reiner Essigsäure und eines Substratgemisches (30% Essigsäure + 70% Saccharose) sind in Tabelle 2 aufgeführt.

Tabelle 2. Werte für m_S und $\alpha_{x_0}^S$ bei der Verhefung von Saccharose, Essigsäure und eines Saccharose-Essigsäure-Gemisches (System: Hefe H2/78; C-Limitation, Chemostat)

Substrate	m_S (g S/g *HTS* · h)	$\alpha_{x_0}^S$ (g S/g *HTS*)	$\alpha_{x_{1,2}}^S$ $t_R = 3$ h (g S/g *HTS*)	$t_R = 10$ h
Saccharose	0,075	1,69	1,92	2,44
Essigsäure	0,392	1,29	2,47	5,21
Saccharose/	0,255	1,13	1,90	3,68
Essigsäure (70/30 Masse-%)	0,170[1]	1,57[1]	2,08[1]	3,27[1]

[1] theoretisch berechnet für den Fall, daß sich die einzelnen C-Substrate bei der parallelen Assimilation nicht gegenseitig beeinflussen. Bilanzgleichung:

$$\alpha_{x1,2}^S = a \cdot \alpha_{x_0}^{S_1} + (1 - a) \cdot \alpha_{x_0}^{S_2} + a \cdot m_1 \cdot t_R + (1 - a)\, m_2 \cdot t_R$$

(S_1 = Saccharose; a = Masseanteil Saccharose = 0,7; S_2 = Essigsäure)

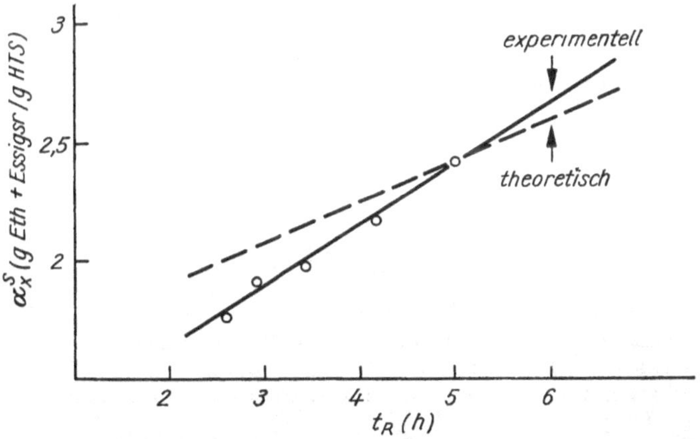

Abb. 5. Abhängigkeit des Verbrauchskoeffizienten für das Substratgemisch Saccharose/Essigsäure (70/30 Masse-%)
Korrektur: *y*-Achse statt Eth. Sacch.

Kurve (——): experimentell ermittelt
Kurve (– – –): kalkuliert aus den α_x^S/t_R-Beziehungen der reinen C-Substrate
(System: Hefe H 2/78; C-Limitation)

Aus den Daten der Tabelle 2 und den Kurven der Abb. 5 ist zu erkennen, daß bei der Verhefung des Substratgemisches von Saccharose/Essigsäure (70/30 Masse-%) unter C-Limitation der Gesamtverbrauchskoeffizient ($\alpha_{x1,2}^S$) im Verweilzeitbereich von $t_R = 2$ bis 5 Std. deutlich unter dem theoretisch zu erwartenden Verbrauchskoeffizient liegt.

Somit kann bei Essigsäurezusatz von anteilig 30 Masse-% zur saccharosehaltigen Nährlösung eine energetisch und stofflich bessere Verwertung der Komponenten im Gemisch gegenüber der Verwertung als Einzelkomponenten erreicht werden.
Bei $t_R > 5$ Std. ist die Effizienz der Fermentation der Einzelkomponenten jedoch zunehmend besser als die des Gemisches der C-Substrate.

Die drei Beispiele zeigen, daß ein minimaler Erhaltungskoeffizient allein noch kein Kriterium für das Erreichen minimaler Verbrauchskoeffizienten darstellt.
Für die optimale Auswahl der Zustandsbedingungen der Fermentation zur Erreichung minimaler Verbrauchskoeffizienten ist die Kenntnis der chemischen Zusammensetzung und die Wirkung der Einzelkomponenten auf die α_x^S/t_R-Abhängigkeit notwendig.

Symbolverzeichnis

BTS	Trockensubstanz für Bakterien
HTS	Trockensubstanz für Hefen
K, K_1, K_2, K_3	allgemeine systemspezifische Konstanten
R	allgemeine Gaskonstante
S_E	C-Substratkonzentration am Fermentoreingang (g/l)
T	Temperatur (Kelvin)
Y	Ausbeutekoeffizient (g trockene Biomasse/g Substrat)
Y_0	Ausbeutekoeffizient bei $t_R = 0$
ΔH	Reaktionsenthalpie (kcal/mol)
C_{H^+}	Wasserstoffionenkonzentration (g/l)
m_S	Erhaltungskoeffizient (g Substrat/g trockene Biomasse · h)
t_R	Verweilzeit (h)
$\dot{x}$	Produktivität (g · BTS/l · h)
α_x^S	Verbrauchskoeffizient (g Substrat/g trockene Biomasse)
$\alpha_{x_0}^S$	Verbrauchskoeffizient bei $t_R = 0$
μ	spezifische Wachstumsrate (h^{-1})

Eingegangen: 4. 1. 1980

Literatur

[1] Pirt, S. J.: Proc. Roy. Soc. Ser. B 163 (1965) 224.
[2] Topiwala, H., Sinclair, C. G.: Biotechnol. Bioengn. 18 (1976) 1827.
[3] Cooney, C. L., Makiguchi, N.: Biotechnol. Bioeng. Symp. Nr. 7 (1977) 65.
[4] Ringpfeil, M., Wünsche, L., Klappach, G.: VII. Intern. Symp. of Continuous Cultivation of Microorganisms, Prag 1978.
[5] Schneider, J., Frommholz, G., Feiler, E., Jechorek, M.: VII. Intern. Symp. of Continuous Cultivation of Microorganisms, Prag 1978.
[6] Edwards, V. H.: Biotechnol. Bioeng. 12 (1970) 679.
[7] Hueting, S., Tempest, D. W.: Arch. Microbiol. **115** (1977) 73.

Acta Biotechnologica 0 (1980) 0, 55–60

Иммобилизация микробного молокосвертывающего фермента на неорганическом носителе

К. А. Калунянц, Л. А. Нахапетян, И. Д. Стальная, В. Н. Борисова

Всесоюсный научно-исследовательский биотехнический институт,
г. Москва, ул. Кропоткинская, 38

Zusammenfassung

Es wird die Immobilisierung eines Milchgerinnungsenzymes (als Präparat) aus der Kultur des niederen Pilzes *Mucor pusillus Lab. sp. renninus* durch Adsorption und kovalente Bindung an einen anorganischen Träger beschrieben. Die Adsorption wurde an Amino- und Karboxylsilochrom und die kovalente Bindung an Azid-Derivaten von Silochrom untersucht. Proben mit einer hohen Aktivität wurden sowohl durch Adsorption (1200 Einheiten/g Präparat) als auch durch kovalente Bindung (890 Einheiten/g Präparat) gewonnen. Die Möglichkeit des Einsatzes von immobilisierten Präparaten zur mehrfachen Gerinnung von Milch wird erörtert. Die stabilsten Präparate wurden durch die Methode der kovalenten Immobilisierung hergestellt.

В последние годы замена дефицитного сычужного фермента молокосвертывающими ферментами микробного происхождения одной из актуальных проблем сыроделия [6]. Однако, до настоящего времени не удается осуществить полную замену сычужного фермента микробными протеазами, так как последние, вызывая неспецифический протеолиз, приводят к появлению нежелательных изменений вкуса при производстве некоторых видов сыров [4].
Использование для свертывания молока в сыроделии иммобилизованных микробных молокосвертывающих ферментов, интерес к которым особенно возрос в последнее десятилетие, открыло бы новые перспективные пути для устранения ряда недостатков, связанных с применением растворимых микробных ферментов, поскольку появляется реальная возможность создания процесса непрерывного свертывания молока и контролируемого выведения иммобилизованного микробного фермента из технологической схемы.
Целью данной работы являлось получение иммобилизованного путем адсорбции и ковалентного связывания микробного молокосвертывающего фермента и изучение возможности его применения для многократного свертывания молока.

Экспериментальная часть

Методика

Для иммобилизации использовали препарат молокосвертывающего фермента, выделенного из культуры плесного гриба *Mucor pusillus Lab. sp. renninus*, с активностью 300000 ед/г препарата. Фермент имеет молекулярный вес 38000–40000 и обладает наибольшей стабильностью в диапазоне pH от 3,0 до 5,5 [10].

Препарат микробного фермента был любезно предоставлен Смирновой Т. А. (МТИПП).

В качестве носителя использовали пористый кремнезем-силохром С-80 (размер частиц 0,3—0,5 мм, размер пор 400—600 Å, удельная поверхность 80 м²/г), модифицированный различными методами с использованием реакций [1, 7, 11]:

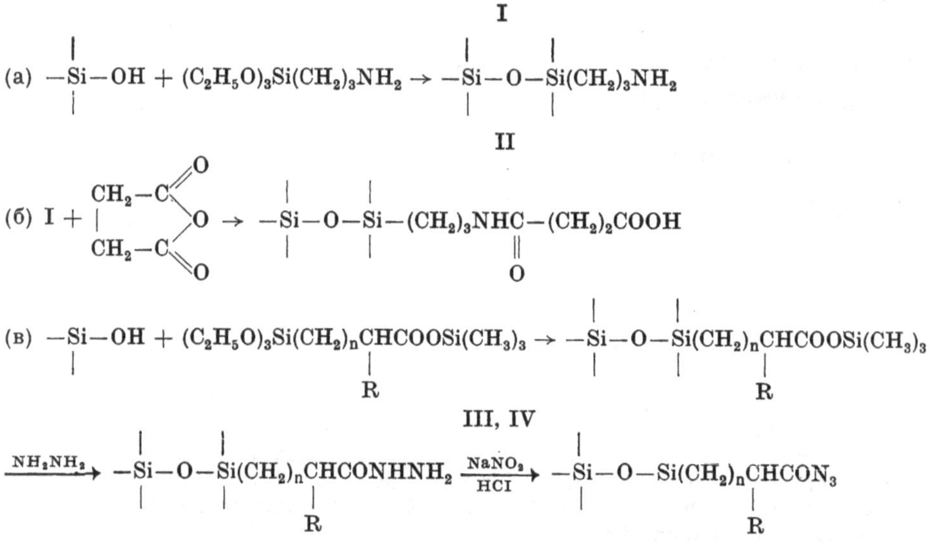

где III $n = 2$, $R = H$, IV $n = 1$, $R = CH_3$

Аминопропилсилохром (I) и сукциниламидопропилсилохром (II) содержали 0,75 мг-экв аминогрупп и 0,64 мг-экв карбоксильных групп на 1 г носителя соответственно. Содержание реакционноспособных групп на носителях III и IV определяли по содержанию азота в устойчивой при хранении гидразидной форме. Концентрация гидразидных гупп составляла 0,22 мг—экв/г носителя.

Адсорбционная иммобилизация микробного молокосвертывающего фермента на носителях I и II

Различные исходные количества нативного фермента (от 20 до 100 мг белка) растворяли в 10 мл 0,05 М универсального буфера (смесь равных объемов фосфорной, уксусной и борной кислот) с соответствующим значением pH, приливали к заданному количеству модифицированного носителя и инкубировали при комнатной температуре при перемешивании в течение 0,5—3,0 часов. Затем иммобилизованный препарат промывали на фильтре водой и 0,05 М ацетатным буфером pH 3,5 до полного отсутствия белка в промывных водах [9].

Иммобилизация ковалентным связыванием микробного молокосвертывающего фермента на носителях III и IV

К 1 г модифицированного носителя добавляли 10 мл универсального буфера с соответствующими значениями pH, содержащего различные исходные количества нативного фермента (от 20 до 100 мг белка), и инкубировали в течение 1—6 часов при 4 °C при перемешивании. Иммобилизованный препарат промывали на фильтре водой 1 М раствором хлористого натрия и 0,05 М ацетатным буфером pH 3,5 до полного отсутствия белка в промывных водах [9].

Молокосвертывающую активность нативного фермента определяли по методу [3]. Молокосвертывающую активность иммобилизованного фермента определяли в условиях проведения ферментативной стадии при 4 °C, а стадии коагуляции при 35 °C по методу [2]. Определение ,,растворимой" активности, которая могла появиться в результате десорбции фермента при контакте с молоком, проводили согласно методу [12]. Свертывающую активность растворимого и иммобилизованного ферментов выражали как величину, обратную времени свертывания молока в стандартных условиях определения (сек$^{-1}\times 10^3$). Белок в растворе нативного фермента и на носителе оределяли по методу Хартри [9].

Результаты и их обсуждение

Основной предпосылкой к использованию аминопропилсилохрома (носитель I) в качестве носителя для адсорбции кислой микробной протеазы послужило наличие положительного заряда на его поверхности при кислых значениях pH. В то же время было интересно использовать носитель с карбоксильной группой (носитель II), так как отрицательный заряд поверхности носителя, по-видимому, должен препятствовать дополнительной адсорбции отрицательно заряженных компонентов молока в процессе длительного контакта.

При изучении адсорбции на носителях I и II исследовали влияние различных факторов, в том числе соотношение фермент-носитель, pH реакционной среды и время реакции.

Иммобилизацию микробного фермента адсорбцией на носителях I и II из буферных растворов проводили в широком интервале значений pH (от 3,0 до 7,0). Выбор диапазона pH обусловлен как pH-стабильностью исследуемого фермента в кислой области (3,0—5,5) [10], так и тем, что последующий ферментативный гидролиз молока иммобилизованными препаратами должен проводиться при pH 6,5—6,7.

Как видно из таблицы 1, лучшие образцы при адсорбции на аминопропилсилохроме были получены при значениях pH 3,0—4,0, при этом молокосвертывающая актив-

Таблица 1. Влияние pH реакционной среды на активность адсорбированного микробного фермента (Количество белка 73 мг, молокосвертывающая активность 2900 ед, 300 мг сухого модифицированного носителя, время реакции 60 минут)

pH	Носитель фунц. группа	Присоединилось к 1 г. носителя		Связалось, % от исходного количества		Уд. ак-ть
		ед. активности	мг белка	активности	белка	ед/мг
3,0	I	298	8,8	10	12	34
4,0	I	209	7,8	7	11	27
5,0	I	122	7,9	4	11	15
6,5	I	38	7,2	1	10	5
3,0	II	1247	16,2	43	22,3	77
4,0	II	833	12,0	28,7	16,4	69
5,0	II	604	11,0	20,8	14,6	55
6,5	II	116	8,3	4,0	11,4	14

ность составляла всего лишь 7—10% от исходной. В противоположность этому при адсорбции микробного фермента на носителе с карбоксильными группами в области pH от 3,0 до 5,0 мы наблюдали значительное, практически двукратное, увеличение удельной активности полученных иммобилизованных препаратов. Как и в случае аминопропилсилохрома наибольшее количество белка сорбировалось при значениях pH 3,0—5,0.

Обнаруженный факт значительного увеличения удельной активности, по-видимому, может быть обусловлен несколькими причинами: влиянием карбоксильных групп носителя, которые могут препятствовать адсорбции отрицательно заряженных пептидов и белков молока, включая казеин [8], избирательной адсорбцией молокосвертывающего фермента по сравнению с другими белками, присутствующими в исходном ферментном препарате, удлинением углеродной цепи модифицирующего агента. Нельзя также исключить и тот факт, что определение активности иммобилизованного фермента проходит в более оптимальных условиях, чем нативного, поскольку известно, что иммобилизация на карбоксилсодержащем носителе может приводить к сдвигу pH-оптимума в щелочную область [5].

В дальнейших опытах при изучении влияния соотношения фермент-носитель адсорбцию на носителях I и II проводили при pH 4,0. Полученные данные (табл. 2) показывают, что с увеличением исходного количества ферментного препарата, вводимого в реакцию, при заданном количестве носителя происходило увеличение как количества адсорбционного белка, так и молокосвертывающей активности, причем для обоих носителей не было достигнуто предела насыщения как по белку, так и по активности.

Оптимальную степень заполнения оценивали по зависимости величины молокосвертывающей активности адсорбированного микробного фермента от количества носителя, вводимого в реакцию, при постоянной концентрации ферментного препарата в реакционной среде (70 мг белка в 10 мл 0,05 M универсального буфера pH 4,0). При этом было найдено, что иммобилизованные препараты с максимальной активностью (300 ед/г и 1200 ед/г для носителей I и II соответственно) получены при адсорбции 70 мг белка на 300 мг сухого модифицированного носителя.

Изменение продолжительности взаимодействия фермент-носитель от 0,5 до 3,0 часов при pH 4,0 показало, что состояние равновесия достигается в течении часа при комнатной температуре.

Таблица 2. Влияние соотношения фермент-носитель на активность адсорбированного микробного фермента (Количество белка 20—100 мг, 300 мг сухого модифицированного носителя, pH 4,0, время реакции 60 минут)

Носитель функц. группа	Добавлено к 1 г носителя		Присоединилось к 1 г носителя		Уд. ак-ть	Связалось, % от исходного количества	
	ед. активности	мг белка	ед. активности	мг белка	ед/г	активности	белка
I	970	25	8	4,3	2	0,8	17
	1900	51	27	5,6	5	1,4	11
	2850	71	200	7,8	26	7,0	11
	3900	100	340	10,8	31	8,7	11
II	970	25	93	2,8	33	10,0	11
	1900	51	323	6,4	51	17,0	12,6
	2850	71	815	12,5	65	29,0	17,6
	3900	100	1940	26,5	73	50,0	26,5

Таблица 3. Влияние pH реакционной среды на активность иммобилизованного ковалентным связыванием на носителе IV микробного фермента
(Количество белка 73 мг, молокосвертывающая активность 2900 ед., 1 г сухого модифицированного носителя, время реакции 60 минут)

pH	Присоединилось к 1 г носителя		Связалось, % от исходного количества		Уд. ак-ть
	ед. активности	мг белка	активности	белка	ед/мг
3,0	870	19,4	30	27	45
4,0	900	17,5	31	24	51
5,5	551	14,6	19	20	38
7,0	189	8,3	7	12	23
8,0	93	6,8	3	9	14

Иммобилизацию микробного фермента на азидных производных силохрома проводили в широком интервале значений pH (от 3,0 до 8,0), поскольку реакция ковалентного связывания полнее протекает в щелочной среде за счет активации N-концевых аминогрупп фермента, в то время как максимальная pH-стабильность фермента наблюдается в дипазоне pH от 3,0 до 5,5 [10]. Исследование влияния pH проводили на носителе IV.

Результаты, полученные при проведении иммобилизации, показали, что реакция лучше проходит при кислых значениях pH (таблица 3). Иммобилизация микробного фермента при pH 3,0—4,0 также сопровождалась увеличением удельной активности полученных препаратов на 10—30% по сравнению с нативным ферментом, что, по видимому, вызвана теми же причинами, что и при адсорбции на карбоксилсодержащем носителе.

При выявлении оптимального количества фермента, вводимого в реакцию (таблица 4) при pH 3,2, было найдено увеличение как количества связанного с носителем белка, так и активности иммобилизованного препарата с увеличением исходного количества ферментного препарата. При этом образцы с более высокой активностью

Таблица 4. Влияние количества фермента на активность микробного препарата, иммобилизованного ковалентным связыванием
(Количество белка от 20 до 100 мг, 1 г сухого модифицированного носителя, pH 3,0, время реакции 60 минут)

Носитель	Добавлено к 1 г носителя		Присоединилось к 1 г носителя		Уд. ак-ть	Связалось, % от исходного количества	
	ед. активности	мг белка	ед. активности	мг белка	ед/мг	активности	белка
III	880	21	24	4,7	5	2,7	22,4
	1800	43	150	8,6	17,4	8,3	20,0
	2800	70	275	10,2	27,0	9,8	14,6
	3760	94	376	13,0	29,0	10,0	13,8
IV	880	21	83	5,8	14,3	9,4	27,6
	1800	43	587	12,5	47	32,6	29,1
	2800	70	868	17,0	51	31,0	24,3
	3760	94	894	19,0	46	19,6	21,7

были получены на слохроме, модифицированном триэтоксисилилизомасляной кислотой.

Препараты с наивысшей молокосвертывающей активностью, полученные адсорбцией на носителе II и ковалентным связыванием на носителе IV, были изучены в процессе многократного свертывания молока. При этом 100 мг препарата иммобилизованного фермента последовательно вводили в реакцию со 100 мл молока при pH 6,5, 4 °C и времени контакта 30 минут при постоянном перемешивании.

После обработки 10 порций молока (общий объем молока 1,0 л) препаратом адсорбированного фермента с активностью 1200 ед/г остаточная активность составила 30% исходной. При этом мы наблюдали появление ,,растворимой'' активности в молочной сыворотке, что, по-видимому, связано с десорбцией фермента в результате длительного контакта с молоком. В то же время препарат, полученный ковалентном связыванием, с активностью 900 ед/г после обработки 23 порций молока (общий объем 2,3 л) имел остаточную активность 50%, то есть практически в два раза выше, чем при использовании препарата адсорбированного микробного фермента. Мы не обнаружили появление ,,растворимой'' активности в молочной сыворотке. Наблюдаемое снижение молокосвертывающей активности, по-видимому, частично связано со стерическими затруднениями, создаваемыми адсорбированными в процессе контакта компонентами молока, поскольку отмывание образцов растворами 2 M хлористого натрия, 0,1 н HCl и 0,2 M ацетатным буфером pH 3,5 приводило к некоторому восстановлению молокосвертывающей активности.

Таким образом, результаты проделанной работы показали, что высокоактивные препараты иммобилизованного микробного фермента могут быть получены как адсорбцией и ковалентным связывамием на носителе с карбоксильной группой. При этом более стабильные препараты получены методом ковалентной иммобилизации.

Eingegangen: 25. 2. 1980

Литература

[1] Борисова В. Н., Стальная И. Д., Меняйлова И. И., Мотина Л. И., Нахапетян Л. А., Федотов Н. С., Рыбалка И. Г.: Прикладная биохимия и микробиология **14** (1978) 5, 709—712.

[2] Долгих Т. В., Суровцев В. М., Козлов Л. В., Антонов В. К., Гинодман Л. М., Звягинцев В. И.: Прикладная биохимия и микробиология **7** (1971) 6, 686.

[3] Евтихов П. Н., Ильина Л. П., Новоселова В. М.: В сб. 2-го Всесоюзного Совещания по ферментам микроорганизмов в г. Минске. 1978. Часть II, 54.

[4] Звягинцев В. И., Гудков А. В.: Мол. пром. 7 (1970), 15—22.

[5] Иммобилизованные ферменты. Современное состояние и перспективы. Под редакцией И. В. Березина, Москва 1976, 2, 60.

[6] Милько А. А., Смирнова Т. А., Балашова Н. А., Никитина Е. Л.: Микробиологич. пром. 4 (1975) 124, 13—15.

[7] Скрябин К. Г., Варламов В. П., Захарьев В. М., Рогожин С. В., Баев А. А.: Биоорг. химия **2** (1976) 10, 1416—1418.

[8] Cheryan M., Van Wyk P., Richardson T., Olson N.: Biotech. Bioeng. **18** (1976) 273.

[9] Hartree G.: Analyt. Biochem. **48** (1972) 422.

[10] Oka T., Ishino K., Tsuzuki H., Morihara K., Arima K.: Agr. Biol. Chem. **37** (1973) 5, 1177.

[11] Robinson P., Dumnill P., Hilly M.: Biochim. Biophys. Acta **242** (1971) 659.

[12] Taylor M., Cheryan M. Richardson T., Olson N.: Biotech. Bioeng. **19** (1977) 883.

Acta Biotechnologica 0 (1980) 0, 61–64

Mischsubstratfermentation – ein energetisch begründetes Konzept

W. Babel

Akademie der Wissenschaften der DDR
Institut für technische Chemie Leipzig
7050 Leipzig, Permoserstr. 15

Summary

On the basis of the combustion enthalpy of microbial biomasses carbon substrates for the aerobically heterotrophic growth of microorganisms can be divided into substrates with energy excess (e.g. ethanol, n-alkanes) and with energy deficit (or carbon excess, e.g. acetate, glucose). By combining substrates having different carbon energy ratios from both groups in a certain mixing proportion an improvement of the yield coefficients linked to a reduction of the specific heat production (kJ/g dry weight), an increase of the growth rate (h^{-1}) and the velocity of the substrate utilization ought to be possible on condition that the co- or auxiliary-substrate does not cause inhibition or (catabolite-) repression effects.

There are some examples corroborating the theoretical concept, e.g. the combinations ethanol/sucrose-*Candida utilis*, methanol/xylose-*Hansenula polymorpha*. In each case the specific growth rate and the yield coefficient on the mixture were higher than on any single substrate of the combination.

Thermochemical methods are suitable for the optimizing the substrate mixing proportion.

Bei der chemo-organo-heterotrophen Ernährung von Mikroorganismen ist die Kohlenstoffquelle zugleich Energiequelle. Im Verlaufe des Stoffwechsels wird die in den Kohlenstoffsubstraten enthaltene Energie in biologisch verfügbare transferiert. Dabei entstehen niedermolekulare Intermediate, die das Kohlenstoffgerüst für die Zellsubstanzsynthese liefern. Auf einem bestimmten Niveau des Katabolismus steht also einem energetisch entwerteten Kohlenstoffskelett die biologisch verwertbare Energie in Form von ATP und/oder Reduktionsäquivalenten gegenüber:

$$\text{Glucose} \rightarrow 2\,\text{Pyruvat} + 2\,\text{NAD(P)H} + 2\,\text{ATP (via Glykolyse)}$$
$$\text{Glycerin} \rightarrow 1\,\text{Pyruvat} + 2\,\text{NAD(P)H} + 1\,\text{ATP}\ [1]$$

Aus derartigen Summengleichungen geht hervor, daß sich die Substrate im Kohlenstoff/Energie-Verhältnis unterscheiden, und es läßt sich auch eine Reihenfolge aufstellen. Im Hinblick auf maximale Ertragskoeffizienten bei der Zellsubstanzsynthese erhebt sich jedoch die Frage nach dem optimalen Kohlenstoff/Energie-Verhältnis. Eine Antwort darauf kann erhalten werden, wenn ein Normierungsmaßstab zur Bewertung der Substrate gefunden wird.

Die Verbrennungsenthalpie der Biomasse ist ein solches Bewertungskriterium. Obwohl die makromolekulare Zusammensetzung (Protein, Kohlenhydrate, Lipide, Nucleinsäuren) der Biomasse variabel ist, unterscheiden sich die Verbrennungsenthalpien verschiedener Bakterien- und Hefespecies bei nicht-limitiertem Wachstum (exponentielle Wachstumsphase) unabhängig von der verwendeten Kohlenstoff- und Energiequelle

praktisch nicht. Gemessen an der Verbrennungsenthalpie der Biomasse lassen sich die Substrate für die chemo-organo-heterotrophe Ernährung bewerten. Danach kann unterschieden werden zwischen Substraten mit Energie-Überschuß und solchen mit Kohlenstoff-Überschuß [2].
Methanol, Ethanol, n-Alkane gehören zur ersten Kategorie, Glucose, Acetat sind Substrate mit Energie-Defizit und Glycerin hat ein nahezu ausbalanciertes Kohlenstoff/Energie-Verhältnis.
Bei Substraten mit Energie-Überschuß sind die höchsten Y_C-Werte (g Trockensubstanz pro g Substrat-Kohlenstoff) zu erwarten. Es sind sogar Werte nahe 100% nicht undenkbar. Im Falle des Methans wäre, wenn 60% der Energie biologisch nutzbar gemacht werden könnten, der Kohlenstoff restlos assimiliert, d. h. in die Zellsubstanz inkorporiert. Tatsächlich liegen aber die experimentellen Y_C-Werte für Substrate mit Energie-Überschuß wie für diejenigen mit einem ausgewogenen Kohlenstoff/Energie-Verhältnis nur bei ca. 68% [3, 2].
Formal kann die Energie von jedem Substrat mit etwa dem gleichen sog. thermodynamischen Wirkungsgrad [4] in ATP transferiert werden [2]. Die experimentellen Y_C-Werte deuten jedoch darauf hin, daß in Abhängigkeit von der Kohlenstoff- und Energiequelle für die chemo-organo-heterotrophe Ernährung der Mikroorganismen bei Substraten mit Energie-Überschuß Energiekonservierungssysteme mit geringerer Effizienz gebildet bzw. benutzt werden [5], so daß die maximalen Y_C-Werte — wie bei den Substraten mit Energie-Defizit — energetisch determiniert sind. Der „*mixed function oxidase*"-Mechanismus für die Primäroxydation des Methans, die bisherigen Kenntnisse über das Energiekonservierungssystem methylotropher Bakterien [6, 7, 8, 9], aber auch die Befunde von GALLO u. AZOULAY [10], wonach beim Wachstum von *Candida tropicalis* auf n-Alkanen und langkettigen Fettsäuren der via β-Oxydation freigesetzte Substratwasserstoff biologisch nicht genutzt werden kann, unterstützen diesen Verdacht.
Daß sich die Y_C-Werte für viele Substrate mit Energie-Überschuß auf einem nahezu einheitlichen Niveau einpegeln [11, 3] legt jedoch andererseits die Vermutung nahe, daß die überschüssige Energie — auch wenn sie in biologisch verwertbare transferiert werden würde — nicht für eine effektivere Inkorporation des Substratkohlenstoffs genutzt werden kann, sondern als Wärme verloren gehen muß, weil es die „Konstruktion" des Finalproduktes Zellsubstanz im Sinne des Äquifinalitätsprinzips nicht erfordert oder erlaubt (unter den Bedingungen eines nicht-limitierten Wachstums, bei dem die Zelle nichts anderes tut als sich reproduziert und keine nennenswerten Reservestoffe akkumuliert). Die nachfolgenden Bilanzgleichungen sollen helfen, die Begrenzung der Y_C-Werte zu erklären.

$$2\ \text{Ethanol} \rightarrow \text{Pyruvat} + CO_2 + 18\ \text{ATP}$$
$$\tfrac{1}{4}\ \text{Hexadecan} \rightarrow \text{Pyruvat} + CO_2 + 17\ \text{ATP}$$
$$\tfrac{1}{4}\ \text{Hexadecan} \rightarrow \text{Pyruvat} + CO_2 + 7\tfrac{3}{4}\ \text{ATP [10]}$$
$$\text{Glycerin} \rightarrow \text{Pyruvat} + 7\ \text{ATP [1]}$$
$$\text{Xylose} \rightarrow \text{Pyruvat} + 2CO_2 + 16\ \text{ATP (via G6P-DH u. 6-PG-DH)}$$
$$\tfrac{3}{5}\ \text{Xylose} \rightarrow \text{Pyruvat} + 4\ \text{ATP}$$
$$\tfrac{1}{2}\ \text{Glucose} \rightarrow \text{Pyruvat} + 4\ \text{ATP}$$
$$2\ \text{Essigsäure} \rightarrow \text{Pyruvat} + CO_2 + 6\ \text{ATP}$$

Diese Gleichungen zeigen, daß — unabhängig von der verfügbaren „biologischen" Energie — nur ein bestimmter Teil des angebotenen Substrat-Kohlenstoffs in Form von Pyruvat, das hier als repräsentativ für ein fiktives Biomassemolekül angesehen wird, Platz findet bzw. für die Zellsubstanzsynthese bereitsteht (und daß wahrscheinlich die dem Pyruvat in der Währung von ATP gegenüberstehende „biologische" Energie darüber entscheidet, zu welchem Prozentsatz das Kohlenstoffskelett in die Zellsubstanz inkorporiert werden kann. Danach scheinen auch Methanol und Glycerin zu den energetisch limitierten Substraten zu gehören).

Durch Mischen von Substraten müßte es möglich sein, überschüssige Energie eines Substrates für die Inkorporation „substratfremden" Kohlenstoffs eines anderen, Auxiliar- oder Co-Substrates zu nutzen, was bei geeigneten Substratkombinationen und -mischungsverhältnissen zur Verbesserung der Ertragskoeffizienten und wegen des damit verbundenen erhöhten energetischen Wirkungsgrades zur Senkung der spezifischen Wärmebildung (Kühlung!) sowie zur Senkung des Energieeintrages bei der aeroben Fermentation führen könnte, und die Geschwindigkeit der Substrat-Utilisation zu erhöhen, vorausgesetzt, das Co-Substrat löst keine Inhibitions- oder (Katabolit-) Repressionseffekte aus. Die chemostatische Kultivierung erscheint dafür besonders prädestiniert. (Utilisation umfaßt Assimilation und Dissimilation von Substraten sowie die Akkumulation von nieder- und makromolekularen Verbindungen).

Für Substrate, die durch Fixierung an einen Akzeptor assimiliert werden oder deren Assimilationssequenz heterotrophe CO_2-Fixierung einschließt, ist dieses Prinzip vorgebildet.

Als Beispiel kann die methylotrophe Ernährung [12, 13] gelten. Durch Zuspeisung von CO_2 bei Methanol assimilierenden Bakterien [14, 15] und Hefen [16] bzw. von Xylose bei Methanol assimilierenden Hefen [17, 18] konnte sowohl eine Erhöhung der spezifischen Wachstumsgeschwindigkeit (s. Tab. 1) als auch der Ertragskoeffizienten festgestellt werden. Die Verbesserung der Ertragskoeffizienten bei diesem Co-Metabolismus (in des Wortes eigenster Bedeutung, vgl. [19, 20]) kann unter der Bedingung, daß $\mu_{S12} > \max \mu_{S1}$ und $\max \mu_{S2}$ ist, nicht einfach auf die Erhöhung der spezifischen Wachstumsgeschwindigkeit zurückgeführt werden [21]. Dagegen läßt sich die erhöhte Wachstumsgeschwindigkeit durchaus auch als Überwindung eines Flaschenhalses im Stoffwechsel interpretieren, sie muß nicht notwendigerweise durch ein Ausbalancieren des Kohlenstoff/Energie-Verhältnisses zustande kommen.

In der Tabelle 1 sind weitere Beispiele für die Erhöhung der Wachstumsgeschwindigkeit bei Verwendung von Substratgemischen zusammengestellt. Bei den Kombinationen Ethanol/Saccharose — *Candida utilis* (E. Feiler, pers. Mitteilung) und Methanol/ Xylose — *Hansenula polymorpha* wurden um ca. 10% höhere Ertragskoeffizienten ermittelt.

Tabelle 1. Maximale spezifische Wachstumsraten von *Hansenula polymorpha* beim Wachstum auf verschiedenen Substraten und Substrat-Kombinationen

Substrat		Substrat-Kombinationen	
		Methanol	Ethanol
Methanol	0,155		
Xylose	0,12	0,21	
Glycerol	0,16	0,225	
Ethanol	0,305		
Fructose	0,275		0,37
Saccharose	0,23		0,315
Glucose	0,41	0,407	0,409

Die Wachstumsgeschwindigkeit μ (h^{-1}) wurde aus batch-Experimenten ermittelt. Bei den einzelnen Substraten wurde eine Startkonzentration von 1 g/l gewählt. Die Mischungsverhältnisse für die verschiedenen Substratkombinationen wurden variiert.

Die Wachstumsgeschwindigkeiten ergaben sich bei einem Mischungsverhältnis von 1:1, dabei erfolgte die Utilisation der Substrate simultan.

Im Falle der Kombination reduzierte C_1-Verbindung/Kohlendioxid haben wir es mit einem Extrem des Co-Metabolismus bei der chemo-organo-heterotrophen Ernährung zu tun, die Art der CO_2-Assimilation könnte ähnlich wie die Assimilation von Methanol via Ribulosediphosphat-Weg [22, 23] sogar als chemo-organo-autotroph charakterisiert werden. Aber gewissermaßen das Extrem dieses Auxiliarsubstrat-Prinzips ist realisiert bei der chemo-litho-autotrophen Ernährung zum Beispiel von „Knallgasbakterien". Das Konzept basiert auf starken Vereinfachungen.
Die Einteilung der Substrate bleibt im Prinzip auch erhalten, wenn der Berechnung die Freien Standardbildungsenthalpien der einzelnen Reaktanten zugrunde gelegt werden, und es gibt Beispiele, die dieses Konzept bestätigen, so daß nunmehr biochemische Fragen, insbesondere nach der Regulation dieser bioenergetischen Prozesse, in den Vordergrund treten.
Aus eigenen Experimenten wissen wir, daß die Kapazität für die Assimilation (immer) kleiner ist als die für die Dissimilation.
Nur wenn eine bestimmte Substratkonzentration nicht überschritten wird, und bei einem bestimmten Mischungsverhältnis kann das Mischen von Substraten zu einer effektiveren Ausnutzung führen. Thermochemische Methoden bieten sich für die Optimierung besonders an.

Eingegangen: 17. 12. 1979

Literatur

[1] Lin, E. C. C.: Ann. Rev. Microbiol. **30** (1976) 535.
[2] Babel, W.: Z. Allg. Mikrobiol. **19** (1979) 671.
[3] Linton, J. D., Stephenson, R. J.: FEMS Microbiology Letters **3** (1978) 95.
[4] Decker, K., Jungermann, K., Thauer, R. K.: Angew. Chem. **82** (1970) 153.
[5] van Verseveld, H. W., Boon, J. P., Stouthamer, A. H.: Arch. Microbiol. **121** (1979) 213.
[6] Colby, J., Dalton, H.: Biochem. J. **157** (1976) 495.
[7] Tonge, G. M., Drozd, J. W., Higgins, I. J.: J. Gen. Microbiol. **99** (1977) 229.
[8] Netrusov, A. J., Rodionov, Y. N., Kondratieva, E. N.: FEBS Letters **76** (1977) 56.
[9] Anthony, C.: J. Gen. Microbiol. **104** (1978) 91.
[10] Gallo, M., Azoulay, E.: Biotechnol. Bioeng. **17** (1975) 1705
[11] Payne, W. J.: Ann. Rev. Microbiol. **24** (1970) 17.
[12] Babel, W.: Dissertation B, Akademie der Wissenschaften der DDR, 1977.
[13] Babel, W.: Wiss. Z. Ernst-Moritz-Arndt-Univ. Greifswald.
[14] Kirikova, N. N., Romanova, A. K.: Mikrobiologiya XLI (1972) 202.
[15] Babij, T., Ralph, B. J., Rickard, P. A. D.: In: Proceedings of the International Symposium on Microbial Growth on C_1-Compunds, p 213. Tokyo, Japan, 1974.
[16] Malashenko, Yu. R., Gagarina, V. A.: XII. Internationaler Kongreß für Mikrobiologie München, 1978. Abstract F 13.
[17] Scholz, R., Hilger, U., Lingner, G., Babel, W.: Vortrag anläßlich der Tagung „Regulation und Bioenergetik der mikrobiellen Substratverwertung" der Sektion Mikrobiologie der Biologischen Gesellschaft der DDR 1978.
[18] Scholz, R., Lingner, G., Hilger, U., Babel, W.: Poster-Demonstration anläßlich der 4. Vortragstagung und Hauptversammlung der Sektion Mikrobiologie der Biologischen Gesellschaft der DDR. Reinhardsbrunn, 1979.
[19] Hulbert, M. H., Krawiec, S.: J. Theor. Biol. **69** (1977) 287.
[20] Perry, J. J.: Microbiol. Rev. **43** (1979) 59.
[21] Stouthamer, A. H.: Intern. Reviews of Biochemistry, series 2, Microbial Biochemistry (Ed. by J. R. Quayle) pp. 1—47. University Park Press Baltimore, 1979.
[22] Anthony, C.: Sci. Prog. Oxf. **62** (1975) 167.
[23] Whittenbury, R., Kelly, D. P.: Microbial Energetics (Ed. by B. A. Haddock and W. A. Hamilton) pp 121—149. Cambridge University Press London, New York, Melbourne, 1977.

Acta Biotechnologica 0 (1980) 0, 65–66

SHORT COMMUNICATIONS

Mechanism of Inactivation of Glucoseoxidase by Hydrogenperoxide

D. KIRSTEIN, F. SCHELLER and P. MOHR

Akademie der Wissenschaften der DDR,
Zentralinstitut für Molekularbiologie,
Bereich Angewandte Enzymologie
1115 Berlin-Buch, DDR

The application of glucoseoxidase (GOD) for preparation of gluconic acid in enzyme reactors [1, 2] or for removing of oxygen or glucose from foodstuffs [3] is restricted by its irreversible product inactivation by hydrogenperoxide. To minimize this process several groups coupled the GOD catalyzed reaction with H_2O_2-decomposition by inorganic catalysts [4] or by catalase [5–7]. Coimmobilization of GOD and catalase was rather effective but catalase is inactivated also [8], therefore H_2O_2 decomposition is not a succesfull method to get a long "working stability". Some hints were given in the last year that the inactivation of flavoenzymes occurs by primary radicalic products of the flavoenzyme catalyzed oxidation by oxygen, especially by the O_2- and OH-radical [9].

The participation of flavin semiquinone and O_2-radicals in the GOD catalyzed reaction was postulated on the basis of the one-electron mechanism with the model substrates furanoin and nitroxid [10]. Since oxidized GOD is less destroyed by H_2O_2 than the reduced form the formation of OH-radicals by the reaction of O_2 with H_2O_2 may be responsible for the desactivation of the working enzyme. Therefore, the addition of the OH scavanger mannitol [11] as well as the O_2 destroying agents superoxide dismutase and Cu^{2+} [11, 12] would have to increase the functional stability of the enzyme provided that these species are involved in the desactivation reaction.

Therefore, we tested different types of procedures to prevent radicalic attack of the GOD molecule.

Materials and methods

The enzymes used were commercial products: Glucoseoxidase (P. notatum) VEB Arzneimittelwerk Dresden, 46 U/mg; catalase (beef liver) Reanal, Budapest, 2000 U/mg; peroxidase (HRP) (horse radish) VEB Arzneimittelwerk Dresden, 32 U/mg; superoxid dismutase (SOD) (beef erythrocytes) Sigma Chemical Co. St. Louis/USA, 2700 U/mg.

Glucose, mannitol, copper-(II)-sulfate, hydroquinone were used as reagent grade chemicals without further purification.

The experiments were performed in a titration vessel (10 ml) at 25°C and pH 5.6 (exception: SOD experiments at pH 7.5) using a Radiometer pH-stat device for measuring the reaction rate (gluconic acid production) [13].

The solutions were aerated by means of a membrane pump; a constant glucose level was held by addition of a concentrated glucose solution (500 g/l).

Results and discussion

The half lives of GOD inactivation are given in table 1 for the different systems studied. Copper ions and superoxidde dismutase do not stabilize the GOD, moreover the stability is decreased. The best possibility of GOD stabilization we found is a consecutive reaction, catalyzed by HRP. We choose hydroquinone as modelsubstrate of this reaction because its oxidation product p-benzoquinone is not accumulated in the reaction mixture but as a GOD-substrate [14] it is recycled to hydroquinone.

Table 1. Half live of GOD inactivation during glucose oxidation by oxygen in several systems

GOD-concentration: 20 mg/l
glucose concentration: 30 g/l

Additional component	concentration [g/l]	$\tau_{1/2}$ [min]
Cu^{2+}	$2.49 \cdot 10^{-2}$	95.3
SOD	$4.50 \cdot 10^{-2}$	103.3
mannitol	182.2	191.0
—		194.5
catalase	$5.43 \cdot 10^{-2}$	368.7
HRP/hydroquinone	$10^{-1}/10$	541.5

The ineffectivity of the O_2- and OH-scavangers underlines that these species are not responsible for the desactivation of the enzyme. This is in accord with the small difference in the redox potentials of the two separate electron transfer reactions of the GOD [15] preventing the generation and stabilization of the radical form of the enzyme in the presence of an excess of glucose. The rather good stability in connection with a consecutive HRP catalyzed reaction of H_2O_2 with a hydroquinone/quinone-pair opens possibilities of prolonged GOD application, e.g. with spacer mediated coupling of quinones to a solid GOD/HRP biocatalyst.

Eingegangen: 20. 3. 1980

References

[1] Kirstein D., Kühn W. and Mohr P.: Die Lebensmittelindustrie (in press).
[2] Kirstein D., Besserdich H. and Kahring E.: Chem. Techn. (in press).
[3] Kirstein D. and Kühn W.: Die Lebensmittelindustrie (in press).
[4] Duvnjak Z. and Lilly M. D.: Biotechn. Bioeng. **18** (1976) 737.
[5] Greenfield P. G., Kittrell J. R. and Laurence R. L.: Analyt. Biochem. **65** (1975) 109.
[6] Bouin J. C., Atallah M. T. and Hultin H. O.: Biochim. Biophys. Acta **438** (1976) 23.
[7] Jaworek D.: in "Insolubilized Enzymes" ed. Salomona M. et al. Raven Press 1974 p. 71.
[8] Buchholz K. and Gödelmann B.: Biotechn. Bioeng. **20** (1978) 1202.
[9] Tramper G.: Oxidation of aza-heterocycles by free and immobilized xanthin oxidase and xanthin dehydrogenase Thesis Agric. Univ. Wageningen 1979.
[10] Chan T. and Bruice C.: J. Amer. Chem. Soc. **99** (1977) 2388.
[11] Halliwell B. and Ahulwalia S.: Biochim. J. **153** (1976) 513.
[12] Mayeda E. and Bard A.: J. Amer. Chem. Soc. **95** (1973) 6223.
[13] Kirstein D.: Die Lebensmittelindustrie **24** (1977) 311.
[14] Schläpfer P., Mindt W. and Racine Ph.: Clin. Chim. Acta **57** (1974) 283.
[15] Stankovich T., Schopfer M. and Massey V.: J. Biol. Chim. **253** (1978) 4971.

Acta Biotechnologica 0 (1980) 0, 67–68

BOOK REVIEWS

WANG, D. I. C.; COONEY, C. L.; DEMAIN, A. L.; DUNNILL, P.; HUMPHREY, A. E.; LILLY, M. D.

"Fermentation & Enzyme Technology"

New York, John Wiley & Sons, 1979

(aus der Reihe "Techniques in Pure and Applied Microbiology" herausgegeben von C.-G. HEDÉN, Karolinska-Institut, Stockholm)

In 14 Kapiteln werden mikrobiologische, biochemische und verfahrenstechnische Grundlagen der Biotechnologie dargestellt. Die Autoren sind sämtlich langjährig auf den ausgewählten Gebieten in Forschung und Lehre tätig. Das Buch ist das schriftliche Konzentrat eines Kurses über Fermentationstechnologie, der unter maßgeblicher Beteiligung der Autoren seit 1962 jährlich von M. I. T. veranstaltet wird.

Dem Studenten der Biotechnologie, der angewandten und technischen Mikrobiologie, der technischen Biochemie oder der Bioingenieurtechnik sowie dem aus Mikrobiologie, Biochemie, Chemie oder Ingenieurwissenschaften hervorgegangenen Praktiker in der mikrobiologischen und biochemischen Industrie wird mit diesem Buch ein sicherer Führer durch die methodischen Grundlagen der Biotechnologie in die Hand gegeben. Die Auswahl der Beispiele ist, trotz der Jugendlichkeit des Gebiets, klassisch zu nennen. Die angeführte Literatur markiert die Fixpunkte einer nunmehr drei Jahrzehnte währenden stürmischen Entwicklung der Biotechnologie.

Der mit biotechnologischer Forschung Beschäftigte wird nicht in gleichem Maße befriedigt sein. Für ihn werden eher die Lücken sichtbar, die die biotechnologische Forschung auf wichtigen Teilgebieten bis jetzt nicht geschlossen hat, so z. B. zur sterilen und unsterilen Prozeßführung, zur Prozeßführung der anaeroben und photosynthetischen Fermentation, zu Prozessen mit pflanzlichen und tierischen Zellen. Er wird neuere Entwicklungen und Theorien vermissen, so z. B. zur Fermentortechnik oder zur Wirkung von Phagen auf den Fermentationsprozeß. Es wird ihm auch auffallen, daß die Autoren die mikrobiologische Technologie ausschließlich methodisch darstellen, die biochemische Technologie jedoch mit Elementen einer beschreibenden Technologie auffüllen. Das Buch stellt sich somit als ein ausgezeichnetes Lehrbuch für den Anfänger und Praktiker dar, weniger als ein Lernbuch das dem Kundigeren auf dem Gebiet der Biotechnologie Anregungen zum Nachdenken über aktuellste Problemstellungen gibt. In der Lehre wird es seinen Platz finden und mit Sicherheit als ein Standardwerk geschätzt und genutzt werden.

M. RINGPFEIL

K. A. Kalunjanz, L. I. Golger

Микробные ферментные препараты

Technologie und Ausrüstung

Moskau, Verlag Lebensmittelindustrie, 1979, (russ.) 304 Seiten, 53 Tabellen, 95 Abbildungen, 149 Literaturangaben

Die Gewinnung von Enzympräparaten auf mikrobieller Basis gewinnt in den letzten Jahren mehr und mehr an praktischer Bedeutung.
Das vorliegende Buch spiegelt die umfangreichen und fortgeschrittenen Erfahrungen der Sowjetunion auf diesem Teilgebiet der Biotechnologie wider.
Nach einem kurzen Abriß über die Entwicklung der wissenschaftlichen Grundlagen und über den Aufbau industrieller Produktionen innerhalb der mikrobiologischen Industrie der UdSSR werden Technologien und Ausrüstungen zur Herstellung von Enzympräparaten ausführlich beschrieben. Zunächst werden die allgemeinen Grundlagen (Aktivitätsbestimmungsmethoden, Mikroorganismen, Nomenklatur der in der UdSSR hergestellten Präparate) vorgestellt. Anschließend werden die technologischen Voraussetzungen und Bedingungen bei der Kultivierung von Mikroorganismen zur Gewinnung von Enzympräparaten mit höchster Aktivität bei der Applikation ausführlich erläutert (zweckmäßigste Nährmedien, Grundlagen der Submers- und Emers-Verfahren, Gewinnung der Enzymlösungen und ihre Aufkonzentrierung, Abtrennung und Konfektionierung der Enzyme, Aufarbeitung der Rückstände). Kurz wird auf das Gebiet der immobilisierten Enzyme eingegangen. Der erste Teil des Buches wird mit der Beschreibung einiger technologischer Schemata zur Gewinnung von Enzympräparaten abgeschlossen.
Im zweiten Teil befassen sich die Autoren sehr ausführlich mit dem gegenwärtigen Stand der bei der industriellen Enzymproduktion eingesetzten Ausrüstungen und Geräte, wobei sowohl sowjetische als auch international anerkannte Typen berücksichtigt wurden. Neben der technischen Beschreibung und der Angabe der möglichen Einsatzgebiete wird Wert auf die Darstellung der erreichbaren Effekte gelegt.
Spezielle Aufmerksamkeit wird den modernsten technischen Lösungen, so den Methoden und Apparaturen der Membranfiltration, der kontinuierlichen fraktionierten Enzymtrennung, der Granulierung und Mikrokapsulierung der Enzympräparate u. a., gewidmet.
Gesonderte Abschnitte befassen sich mit den Ausrüstungen zur Vorbereitung der Nährmedien, den Apparaturen zur Submers- und Emerskultivierung der Mikroorganismen, den Ausrüstungen zur Abtrennung der Biomasse, zur Konzentration und Reinigung der Enzymlösungen, den Trocknungsanlagen, den Apparaten zur Konfektionierung der Enzympräparate sowie mit der Beschreibung kompletter technologischer Linien und der Behandlung von Fragen der Sicherheitstechnik. Durch seinen gut gegliederten und übersichtlichen Aufbau und die klare und verständliche Darstellung ist das Buch besonders für Studierende (Fachrichtungen Biotechnologie, Lebensmitteltechnologie, Biowissenschaften u. a.) und sowohl für Forscher als auch für Praktiker in Instituten (Grundlagen- und angewandte Forschung) und Betrieben (Projektierung, Versuchs- und Produktionsanlagen, Chemieanlagenbau) geeignet. Sehr wertvoll kann es für die Fachleute sein, die ingenieur- und verfahrenstechnisch für das in stürmischer Entwicklung begriffene Gebiet der Biotechnologie tätig sind.

W. Kauruff

Contents

FSC
www.fsc.org
MIX
Papier | Fördert
gute Waldnutzung
FSC® C083411

Kolibri 360 GmbH
Ferdinand-Jühlke-Straße 7
99095 Erfurt, Deutschland
produktsicherheit@kolibri360.de